MODIFICATION OF THE INFORMATION CONTENT OF PLANT CELLS

Modification of the Information Content of Plant Cells

PROCEEDINGS OF THE SECOND JOHN INNES SYMPOSIUM
HELD IN NORWICH, JULY 1974.

EDITORS: Roy Markham
D.R. Davies
D.A. Hopwood
R.W. Horne

Symposium Secretary: Miss Justine Speed

1975

NORTH-HOLLAND PUBLISHING COMPANY – AMSTERDAM • OXFORD
AMERICAN ELSEVIER PUBLISHING COMPANY, INC. – NEW YORK

Library of Congress Catalog Card Number: 74 81326

North-Holland ISBN: 0 7204 4505 1
American Elsevier ISBN: 0 444 10699 5

Published by:

North-Holland Publishing Company – Amsterdam
North-Holland Publishing Company, Ltd. – Oxford

Sole distributors for the U.S.A. and Canada:

American Elsevier Publishing Company, Inc.
52 Vanderbilt Avenue
New York, N.Y. 10017

Second John Innes Symposium.
Social Organisation. Mrs. Margaret Markham
                      Mrs. Frances Davies
                      Mrs. Joyce Hopwood
                      Mrs. Doris Horne
                      Miss. Justine Speed

Printed in The Netherlands

CONTENTS

FOREWORD

The Second John Innes Symposium "Modification of the Information Content of Plant Cells" was held at the John Innes Institute, Norwich, on July 16th, 17th, 18th and 19th, 1974. It was attended by 150 visiting scientists from 19 different countries, as well as by a number of the Institute Staff. For the first time the new Conference Hall in the Recreation Centre was used to hold the meetings.

The Seventh Bateson Memorial Lecture "Alternatives to sex; genetics by means of somatic cells" was given by Professor Guido Pontecorvo, F.R.S., on Tuesday evening 16th July, to a large audience. Among them were members of the British Council's Plant Pathology Course who had travelled from Cambridge to be there. Professor Pontecorvo's lecture is published in this volume.

Professor Herbert Stern, who was Chairman of the Session on Transformation, was asked if we might include his introductory remarks in the published Proceedings because of their special relevance to the interesting problems involved. He has kindly provided us with the manuscript reproduced in this book.

The organisation of the Symposium was carried out by the Director, Heads of Departments of the Institute and the Symposium Secretary. Members of the Institute Staff helped in many vital ways during the meeting. The edited manuscripts for this book were prepared in final form by Mrs Patricia Phillips.

This year the programme was arranged to give participants and accompanying family members an opportunity to see something of the City of Norwich. The programme for family accompanying participants included a tour of the Institute buildings and glasshouses, as well as a day spent visiting places of interest in the County of Norfolk.

THE BATESON LECTURE 1974
"Alternatives to sex":   genetics by means of somatic cells.

by

G. PONTECORVO

Imperial Cancer Research Fund Laboratories,
Lincoln's Inn Fields, London WC2A 3PX.

William Bateson, whose memory we are here to honour, put Mendelian genetics on the map in this country and contributed decisively to its early developments. Among other things, he coined the word "Genetics" to designate the study of the "physiology of descent", and he discovered linkage.

By a fortunate coincidence, the title of my lecture honours the memory of another great John Innes man: J.B.S. Haldane (1955). He used the expression "alternatives to sex" to include the approach developed in Glasgow during the previous few years. These two John Innes men stood at the dawn of two eras of genetics: that of genetics *via* sexual reproduction and that of genetics *via* "alternatives to sex".

The field, now fashionably called "somatic cell genetics", includes two distinct, though complementary and overlapping approaches. One is the use of somatic cells in culture, very much like the use of cultures of microorganisms, for the study of general problems: mutation, replication, differentiation, metabolism of macromolecules, selection, social behaviour etc. (reviews: Ephrussi, 1972; Harris, 1970; Puck, 1972; Stoker, 1972). The other is the use of somatic cells, mainly in culture, as an alternative to germ cells for the genetic analysis of the individual from which the somatic cells were taken. In this talk I shall limit myself exclusively to this second field. Let me try first to summarize how it all started.

THE INITIAL WORK WITH ASPERGILLUS NIDULANS.

The initial approach, about 1952, in Glasgow made use of the mould *Aspergillus nidulans*. The formal genetics of this homothallic Ascomycete had been worked out in considerable detail in the previous 5 years by standard techniques based on sexual reproduction. Then my colleague J.A. Roper (1952) developed a simple way of synthesizing strains carrying in their vegetative cells (I shall call them "somatic") diploid nuclei heterozygous for any desired markers. The wild-type strains have haploid nuclei. It became therefore possible to ask the question: does some form of segregation and recombination occur in these diploid

1

"somatic" cells? The idea came to me from the work of 15 years earlier by Curt Stern (1936) on somatic crossing over in *Drosophila*.

After finding that indeed somatic segregation and recombination did occur, though at a low frequency, it took three years to identify the two processes at work and to forge them into efficient tools for routine genetic analysis. Anticipating, it turned out that one of them, somatic crossing over à la Stern, had very limited use as a tool. The other process - involving chromosome loss - turned out to be the basis for a substantial advance in the methods of formal gene-tics (Pontecorvo *et al.*, 1954; Pontecorvo, 1958; Forbes, 1959; McCully and Forbes, 1965). Since 1968 methods identical in principle to those developed for *Aspergillus* are in routine use for human gene-tic analysis by means of somatic cell hybrids (review: Ruddle, 1972).

The possibility of cross-checking the results of analysis based on what came to be called "parasexuality" with the previous ones obtained *via* sexual reproduction was, of course, the great asset of *Aspergillus*. This is why the work was so conclusive. There are now at least 20 times as many workers using *Aspergillus* as there were in those early days: the only part of the work shown to be wrong is a minor hasty cytological conclusion of my own.

The techniques developed then were soon applied to other moulds, including some producing antibiotics. In some of these species sexual reproduction was not known to occur and the "alternatives to sex" were the only way of doing genetic analysis. The first of these asexual species to be analysed was *Aspergillus niger* in 1954.

More important, I became convinced that these alternatives should be usable in somatic cell cultures of higher organisms. In some of these, for example: man, the use of the results of sexual reproduction for genetic analysis is hopelessly restricted by ethical and practical reasons. Clearly a break-through in human genetics could only come from methods by-passing sexual reproduction.

When I first proposed this in the mid-50's and more boldly and pre-cisely at the Ciba Symposium of 1958 (Pontecorvo, 1959), with few ex-ceptions professional human geneticists were not enthusiastic. I had no choice but to try on my own. This was 1956. The U.S. National Institutes of Health and later the M.R.C. gave a generous hand. Un-fortunately my initial line of attack was a failure, mainly because some of the ancillary techniques were not ripe then. They are now, and as you will hear, I have taken up again that initial approach.

Much later, about 1968, the general approach worked out for *Aspergillus nidulans* in the 50's became routine, *mutatis mutandis*, for human genetics. This was the consequence of an utterly unexpected development: the discovery by Weiss and Green (1967) that mouse/man hybrid somatic cells eliminate the human chromosomes on prolonged culturing. The results are spectacular. In the last 6 years - in fact mainly in the last 3 years - about 9 times as many human genes have been assigned to chromosomes or linkage groups as in the whole previous work in human genetics. Unfortunately, quantity is beginning to depress quality.

Table 1

Number of loci assigned since 1968 to
individual human autosomes (June, 1974)

Family studies*	Somatic Cell hybridization	*In situ* annealing	Total
25	61	6	92

* Mainly based on linkage to one or more other loci assigned
by means of somatic cell hybridization.

Most of the young, able and enthusiastic workers in this field use the *Aspergillus* methodology without being aware of where they got it from. This methodology is now beginning to move back to the world of plants where it originated.

GENETIC ANALYSIS BY MEANS OF SOMATIC CELLS.

Genetic analysis means the resolution of the genetic material into its component elements. The techniques for genetic analysis vary tremendously in resolving power, especially between one organism and another. In viruses we consider poor any analysis short of the recognition of nucleotide sequences. In man we are happy if we can identify a gene by its specific effect on the phenotype and assign it to a chromosome or to a linkage group. We are even happier if we can determine the chromosome arm, and exceedingly happy if we can place that gene in its correct order relative to other genes or identify the chromosome band where it lies.

In man resolution almost down to nucleotide sequences has been achieved only in four very special cases all based on *in situ* annealing techniques (review: Hirschhorn and Boyer, 1974). Three involve repetitive sequences: the satellite-DNAs, the 5S ribosomal RNAs and the ribosomal RNAs. One involves the haemoglobin α and β messengers, which are almost the only messengers that reticulocytes produce. The beautiful work of Hirschhorn and his colleagues based on *in situ* annealing of labelled messenger, has led to the location of the two relevant haemoglobin genes on chromosomes 2 and 5 (or 6). Hirschhorn's results are disputed on the grounds that they are physically impossible. H.J. Muller once told me that the physicists had to correct their calibration of an X-ray tube when his mutation results disagreed with their previous calibration. The carbon dating technique is another recent example.

Granted that the ultimate aim in genetic analysis is the resolution of nucleotide sequences and the identification of the active macromolecular products encoded in them, for the present in higher organisms we must be contented with much coarser successive steps. These are: (1) THE IDENTIFICATION OF A GENE by its phenotypic effect and its segregation; (2) THE RECOMBINATION OF THAT GENE with others, syntenic (Renwick, 1971), i.e. carried in the same chromosome pair, or not. (3) THE LOCATION OF THAT GENE within the linear order of other syntenic genes, and (4) THE IDENTIFICATION OF THE CHROMOSOME ARM AND BAND where that gene is physically located.

In classical genetics all of these steps required the use of sexual reproduction. Segregation at meiosis and complementation in heterozygotes permitted to implicate a gene as the cause of a difference in phenotype. Recombination at meiosis was the basis of the two other steps. Cytogenetical techniques, using appropriate structural heterozygotes, were the basis for the physical location.

The work with heterozygous diploid strains of *Aspergillus* showed that all of these operations could be done with "somatic" cells, taking advantage of the two rare processes already mentioned: somatic crossing-over and chromosome loss.

The consequence of somatic crossing-over in a chromosome pair is to produce, from one heterozygous nucleus, two diploid daughter nuclei homozygous in certain precise respects. One is homozygous for all the alleles of one member of that chromosome pair at the loci distal to the position of crossing-over, the other is homozygous at the same loci for the alleles of the other member of that chromosome pair. The linear order of loci can be established by classifying, and an estimate of "distances" made by the relative frequencies of a sample of segregants. Since multiple exchanges are rare enough to be disregarded, the positions of three loci - centromere - A - B - C, in this order - can be established by the fact that homozygosis at A always goes with homozygosis at B and C and homozygosis at B always goes with homozygosis at C.

Differently from *Drosophila*, in *Aspergillus* it was possible to isolate and grow the segregants produced by somatic crossing-over and analyze them by crosses. The results entirely confirmed Stern's deductions of 15 years earlier on the modalities of the process. Unfortunately, for a number of technical reasons, somatic crossing over turned out to be of limited use in formal genetics by means of somatic cells.

The other rare process discovered in *Aspergillus* is quite a different matter. Even in that species it made the assignment of a gene to a linkage group and the detection of synteny and of chromosome rearrangements (e.g. translocations) much simpler than by traditional techniques. But its real value was that it suggested a way of assigning a gene to its chromosome and detecting synteny between two or more genes in higher organisms by means of somatic cells.

This process was misnamed "haploidization" because one, its most obvious, consequence is to produce haploid from diploid somatic cells. It consists in a maldistribution of chromosomes at mitosis, usually affecting one or a few chromosomes at a time. Each maldistribution produces mono-or trisomy. When repeatedly occurring, with strong selection against the inbalanced nuclei, the ultimate result is a population of mainly haploid nuclei reassorting the members of different chromosome pairs in all possible ways.

Since this is a process which involves whole chromosomes, not chromosome segments, in any one haploid clone the alleles at different loci on the same member of a chromosome pair are all present or all absent. In other words in respect of syntenic loci only the parental associations are present among the haploids. The alleles at two non syntenic loci, on the other hand, appear at random in any one of the four possible associations, two parental and two recombinant, in individual haploid clones. A recent example of the application of this rationale to man-Chinese hamster hybrids is given by Van Someren *et al.*(1974) in respect of three syntenic loci on human chromosome 6. Their results are in Table 2. The numbers in the 2 x 2 tables are those of the primary clones showing the presence of either one (+-), or the other (-+), or both (++) or neither (--) of the three enzymes taken two by two at a time.

Table 2
Synteny between human enzymes
PGM$_3$, ME$_1$ and IPO-B in man x Chinese hamster
somatic cell hybrids (Van Someren *et al.*, 1974).

	PGM$_3$ +	PGM$_3$ −		IPO-B +	IPO-B −		PGM$_3$ +	PGM$_3$ −
ME$_1$ +	11	0	ME$_1$ +	18	0	IPO-B +	12	1
ME$_1$ −	0	16	ME$_1$ −	1	14	IPO-B −	0	16

The assignment of a locus to a particular linkage group could be done in *Aspergillus* on the same reasoning as the detection of synteny. But it required the use of genetically multiply marked strains because in this species the individual chromosomes are too small to be identifiable. A diploid, synthesized between a multiply marked strain and a strain carrying a non-assigned gene, produces a number of haploid clones carrying the markers and the non-assigned gene in an array of associations. The non-assigned gene will appear only in the parental associations with the marker (or markers) of the chromosome pair on which it is located.

In a nutshell, chromosome loss provides a very efficient tool because two syntenic loci give zero recombination, while two non-syntenic loci give 50% recombination.

The technique of chromosome loss for detecting synteny and assigning loci became even more efficient after two further technical advances. One was Morpurgo's (1961) discovery that parafluorophenylalanine enormously increases the rate of segregation; followed by Lhoas (1961) demonstration that it acts on haploidization. The other was the routine construction and use of "tester" strains with at least one marker per identified linkage group (Forbes, 1959; McCully and Forbes, 1965).

The rationale of this chromosome loss technique, applied to man/ mouse somatic cell hybrids since Weiss and Green's discovery of 1967 has changed the pace of human formal genetics (Table 1). Instead of using parafluorophenylalanine on heterozygous diploids, one uses human somatic cells hybridised to mouse or Chinese hamster cells. The hybrid system acts on human chromosomes like parafluorophenylalanine acts on *Aspergillus* chromosomes: the human chromosomes are lost one or a few

at a time and the rodent chromosomes retained. Two human loci are
syntenic if they are either lost or retained together in a series of
hybrid clones.

The assignment of a gene to its chromosome does not necessarily require in man the use of genetic markers and tester strains. Human
chromosomes, differently from *Aspergillus*, can be identified microscopically by their morphology and banding patterns. Thus in man/rodent
hybrid cells one can assign a gene to its chromosome by finding, in
a series of hybrid clones, which human chromosome has to be retained
in order that the hybrid cells may express the phenotype determined
by the human unassigned gene.

INITIAL ATTEMPTS WITH HETEROZYGOUS HUMAN CELLS.

In 1956 I took up the challenge of trying to do with heterozygous
diploid human somatic cells what my colleagues and I had done with
Aspergillus heterozygous diploid "somatic" cells. The design was to
grow human skin fibroblasts from individuals heterozygous at a number
of loci expressed in cultured cells. If chromosome loss or non-dis-
junction occurred, it was hoped that clones should arise either fully
haploid, or hemi- or homo-zygous at one or more of the loci.

No human marker usable for selection was available at that time
for the isolation of the expected segregants. Thus it was imperative
to find some way of inducing a manageable rate of chromosome loss.
The various treatments tried - obviously, parafluorophenylalanine was
the first - were ineffective. It also became apparent that aneuploid
human clones were unlikely to survive for long enough until the attain-
ment of a new balance.

This old approach I have know taken up again. The time seems to be
ripe for three reasons. First, there is now an agent which induces
human chromosome loss. Second some of the human inborn errors of
metabolism are now known to be usable as selective markers. Third,
the use of inactivated Sendai virus helps in the fusion of human cells.
These may be from two individuals differing in genetic characters expressed in cell cultures, or from an individual known to be heterozygous for suitable characters.

As to the first reason, De Carli and his colleagues (Rossi *et al.*,
1971; De Carli, 1972) in Pavia showed three years ago that a short
treatment of human cultured cells with the antimycotic griseofulvin
causes a tremendous scatter of chromosome counts from over tetraploid
to subdiploid. A couple of weeks after the treatment mostly tetra-

ploid or diploid counts are found in the cultures. My colleague
Dr. Riddle and I have followed with time-lapse microcinematography what
happens to human skin fibroblasts after griseofulvin treatment. There
is violent elimination of packets of chromosomes from cells in mitosis
and utter disruption of the spindle: hence the formation of cells
with such a scatter of chromosome counts. Clearly, griseofulvin seems
to be what we need.

As to selective markers, there are at least three errors of metabolism in man - Lesch-Nyan syndrome, galactosaemia and *xeroderma pigmentosum* - which lend themselves to selection. Finally the possibility
of producing tetraploids, by fusion of cells from either the same or
different donors, may be of valuable help in reducing the inbalance
of aneuploid clones on their way to become homozygous. Martin and
Sprague, by using cells heterozygous for a chromosomal marker, have
shown a few years ago that indeed human tetraploid cultured cells are
suitable for segregation analysis.

Why bother about pure human somatic cells, when the man/rodent
hybrids have been and are likely to go on being such a wonderful tool?
The answer is simply that the estimated 1000 or so human genes suitable
for mapping by means of the rodent/man hybrids (or indeed other hybrids
which may be synthesized in the future) are prevalently of a special
class. They are mainly "inbuilt markers", in Ephrussi's expression,
i.e. genes determining *inter*specific differences, such as, typically,
a species difference in the electrophoretic form of an enzyme. In
human genetics we are interested in *intra*specific differences, those
that make each individual unique. Only about 1/3 of all these "inbuilt markers" will be polymorphic in man. Anyway, most of the genes
in man which have more immediate interest are known because of the
pathological effects of their recessive alleles, and recessive alleles
do not lend themselves to mapping by the rodent/man hybrid cell technique.

Don't misunderstand me. I do not wish to belittle what we can
still get from the rodent/man hybrid technique. One thousand genes
on the human map, even of a kind that has little direct interest for
bread-and-butter human genetics, will be of enormous value as a frame
for all the other, in my view, more interesting genes. It will tremendously speed up the mapping of these other, even indirectly by
helping the analysis by means of family studies. Finally, it may reveal features of the gene maps and of gene interactions which we can-

not even guess.

Nevertheless, I am firmly convinced that an important complementary pathway in the use of somatic cells for human genetics will be that of using explants from heterozygotes, or synthesizing tetrapoloid hetero- zygotes by fusion between cells from homozygotes.

And now that I have dealt with the prehistory, and guessed at the rosy future, of genetics *via* somatic cells, let me deal with a few salient points of what is going on now.

CURRENT HIGHLIGHTS.

After the initial discovery in 1967 by Weiss and Green of the loss of human chromosomes in man/mouse somatic cell hybrids, three most useful technical advances have followed. One, from Bodmer's group (Nabholz *et al.*, 1969), was the use of human blood white cells as the human component in man/mouse, and later man/Chinese hamster hybrids. This made it possible to make hybrids with the somatic cells of any donor, without the need of selective markers in them.

The second advance was the use for hybridisation of cells from in- dividuals carrying translocations or deletions. This made it possi- ble to go further than merely assigning genes to chromosomes: they may be assigned to chromosome arms and bands within one arm.

The third was, of course, the spectacular advance in cytological techniques, which are improving every month. It is now possible to identify each of the 23 human chromosomes by its banding pattern.

There has also been a promising - though by no means complete - joining of forces between the "hybridizers" and the "family mappers". Any one gene recognizable both in cultured cells and the whole indivi- dual, assigned to a chromosome or to a linkage group by means of one method, can be used to determine synteny in respect of any other gene mapped by the other method. The "somatic cell" and "family" maps are gradually coalescing. For example, on human chromosome 1, of the 18 loci assigned to it, 7 came from somatic cell mapping, 6 from "family" mapping, 4 from both and one from *in situ* annealing and auto- radiography of cDNA.

One result of somatic cell hybridization of considerable interest also for medical genetics is the ascertainment of whether two or more indistinguishable recessive conditions, occurring in different indivi- duals or families, are determined by the same or different genes. These tests of complementation can be carried out by fusing cells from two affected individuals and examining heterokaryotic or hybrid cells for their phenotype. Tests of this kind have added among others *xeroderma pigmentosum* (De Weerd-Kastelein *et al.*, 1972) to the list of genetical disorders in which the same clinical phenotype may result from mutation at different loci.

This leads me to a point that supports the well known, but not endearing, belief of geneticists that the spade work for many biochemical problems is often much more easily done with simple genetical techniques than with rigorous biochemical ones. For example, it is of considerable general interest to know the proportions of monomeric, dimeric etc. enzymes in a number of species. A chemical answer requires the complete purification of each enzyme and its analysis: a Sisyphaean task. Electrophoresis of crude cell extracts gives a first approximate estimate. Cells from heterozygotes for two electrophoretically different forms of an enzyme, or hybrid somatic cells between two species differing in the electrophoretic mobility of an enzyme show in all cases so far both forms of the enzyme. In addition, in a majority of cases they show one or more new electrophoretic bands. These are inferred to be - and in a few cases they have been shown to be - molecular hybrid forms originated by association of polypeptide chains of the two kinds. Association occurs in the cases in which the active enzyme is a polymer. Thus an enzyme can be provisionally classified as monomeric, dimeric etc. on the basis of the number of bands, in addition to the original ones, which are shown in heterozygotes or in interspecific hybrids. Professor Harry Harris has permitted me to show his data based on a sample of 50 human enzymes, some from human heterozygotes and some from man/rodent hybrids. This spade work already tells us that about 1 in 5 human enzymes are monomeric, a matter of considerable general interest. One question to which there should be already an answer is whether or not enzyme polymorphism in human populations is more common in respect of monomeric or of polymeric enzymes. The population genetics of the two kinds of enzymes is likely to be different.

I mentioned that in *Aspergillus nidulans* the use of somatic crossing-over did not play as important a role as was initially hoped in sequencing genes by means of somatic cells. In the first place it required the use of a selective marker placed at the distal end of each chromosome arm. These were hard to come by. In the second place, in that species, sequencing of genes can be done very easily by means of sexual reproduction.

Table 3

Number of polypeptide chains, inferred from heterozygotes
or interspecific somatic hybrids, in 50 human enzymes

MONOMERIC	HOMOPOLYMERIC		
	DIMERIC	TRIMERIC	TETRAMERIC
13	26	1	10

(courtesy of Professor H. Harris, Galton Laboratory)

In man, sequencing of genes by family studies is an extremely slow
process. Any help from other quarters would be most welcome. Unfortunately somatic crossing is not known to occur in human somatic cells
of normal individuals. The beautiful cytological work of German and
his collaborators (German and Pugliatti Crippa, 1966) on cells of individuals affected by Bloom's syndrome, the consequence of a very rare
recessive mutant, shows a tremendous amount of sister chromatid exchanges and what may be interpreted as exchanges at homologous positions
between homologues. The technique used is a very ingenious one based
on the fact that bromodeoxyuridine incorporation in the chromosomes
reduces their stainability with Giemsa or fluorescent dyes. A pulse
of bromodeoxyuridine followed by staining after one or two chromosome
replications shows semiconservative replication much more simply than
autoradiography. This technique applied by German (personal communication) to normal cells shows, after two replications, one chromatid
of each chromosome intensely stained and one much less so. Applied
to Bloom's disease cells it shows the two chromatids as two complementary sequences of intensely stained and poorly stained segments.
The challenge is now either to devise conditions or treatments (mitomycin C has been suggested by German and La Rock (1969)) which will
induce something similar to Bloom's disease in normal cultured cells,
or to introduce Bloom's allele into them. In somatic hybrids between
Bloom cells and rodent cells the effect of Bloom's gene is, as expected, recessive.

There are so many areas of somatic cell genetics I do not know
what else to choose.

I shall finish with two minor contributions of mine. One is the
possibility of predetermining which of the two parental chromosome
complements should be eliminated in hybrid cells. This can be done
by irradiation, or by sensitization with bromodeoxyuridine, of the

cells of one "parent" just before fusion. It can be applied both to inter- and intra-specific hybrids (Pontecorvo, 1974).

The other is the use of the repeated backcross. The backcross (A crossed to B and the F_1 back to A, or B) has long been one of the most powerful tools of genetics *via* sexual reproduction. It may become equally useful in genetics *via* somatic cells, for instance for introducing in cultured somatic cells a gene required for a particular purpose, such as Bloom's syndrome gene.

Since this is a gathering of people interested in plant cells I could not finish without mentioning the momentous experiment of Carlson (1972) and his colleagues who produced whole hybrid plants from the fusion of protoplasts of *Nicotiana glauca* and *N. langsdorfii*. I know that there are doubts about the nature of the results. But such doubts cannot belittle the fact that this was just the kind of experiment which had to be done. The additional examples to be reported at this Symposium by Melchers and Gamborg respectively, confirm that the breakthrough has taken place. While in the case of animals somatic cell fusion is, fortunately, only an analytical tool, in the case of plants - with their totipotentiality of somatic cells - it is also a synthetic tool with fantastic possibilities. What I regret is that most of the workers in this much younger field are unwilling to borrow as much as possible from the much wider experience of their colleagues in the animal field. Since at different times I have had a foot in both and learned from both, I hope this encouragement will not be taken amiss.

The excitement and bubbling activity of the then new Mendelian genetics at the time of Bateson are now repeated with the "alternatives to sex". Bateson and Haldane would have enjoyed the present time tremendously.

REFERENCES

Carlson, P.S., Smith, H.H., and Dearing, R.D. (1972). Parasexual interspecific plant hybridization. *Proceedings of the National Academy of Sciences, U.S.A.* 69, 2292-2294.

De Carli, L. (1972). Personal communication.

De Weerd-Kastelein, E.A., Keijzer, W., and Bootsma, D. (1972). Genetic heterogeneity of *xeroderma pigmentosum* demonstrated by somatic cell hybridization. *Nature New Biology.* 238, 80-83.

Ephrussi, B. *Hybridization of somatic cells.* Princeton University Press. (1972).

Forbes, E. (1959). Use of mitotic segregation for assigning genes to linkage groups in *Aspergillus nidulans*. *Genetical Research*. 6, 352-359.

German, J., and Pugliatti Crippa, L. (1966). Chromosome breakage in diploid cell lines from Bloom's and Fanconi's Anemia. *Annals de Génétique* (Paris). 9, 143-154.

German, J., and La Rock, J. (1969). Chromosomal effects of mitomycin, a potential recombinogen in mammalian cell genetics. *Texas Reports on Biology and Medicine*. 27, 409-418.

Haldane, J.B.S. (1955). Some alternatives to sex. *New Biology*. 19, 7-26.

Harris, Henry. *Cell Fusion*. Harvard University Press, (1970).

Hirschhorn, K., and Boyer, S. (1974). Report of Committee on *in situ* hybridization. *Cytogenetics and Cell Genetics*. 13, 55-57.

Lhoas, P. (1961). Mitotic haploidization by treatment of *Aspergillus niger* diploids with para-fluorophenylalanine. *Nature*. 190, 744.

McCully, K.S., and Forbes, E. (1965). The use of p-fluorophenylalanine with "master strains" of *Aspergillus nidulans* for assigning genes to linkage groups. *Genetical Research*. 6, 352-359.

Martin, G.M., and Sprague, C.A. (1969). Parasexual cycle in Cultivated Human Somatic Cells. *Science*. 166, 761-763.

Morpurgo, G. (1961). Somatic segregation induced by p-fluorophenylala-nine. *Aspergillus News Letter*. 2, 10.

Nabholz, M., Miggiano, V., and Bodmer, W. (1969). Genetic analysis with human-mouse somatic cell hybrids. *Nature*. 223, 358-363.

Pontecorvo, G., Tarr-Gloor, E., and Forbes, E. (1954). Analysis of mitotic recombination in *Aspergillus nidulans*. *Journal of Genetics*. 52, 226-237.

Pontecorvo, G. *Trends in Genetic Analysis*. Columbia University Press, New York, 1958.

Pontecorvo, G. (1959). Panel discussion. pp. 279-285 in *Biochemistry of Human Genetics, Ciba Symposium* (Ed. G.E.W. Wolstenholme and Cecilia M. O'Connor) London, 1959.

Pontecorvo, G. (1974). Induced chromosome elimination in hybrid cells, pp. 65-68 in *Somatic Cell Hybridization* (Ed. R.L. Davidson and F. de la Crux) New York, 1974.

Puck, T.T. *The Mammalian Cell as a Microorganism*. London, Holden-Day, 1972.

Renwick, J.H. (1971). The mapping of human chromosomes. *Annual Review of Genetics*. 5, 81-120.

Roper, J.A. (1952). Production of heterozygous diploids in filamentous
fungi. *Experientia*. 8, 14-15.

Rossi, C., Tredici, F. and De Carli, L. (1971). Effetto della griseo-
fulvina sulla dinamica degli assetti cromosomici della linea umana
eteroploide EUE (abstract). *Atti Associazione Genetica Italiana*.
16, 44-45.

Ruddle, F.H. (1972). Linkage analysis using somatic cell hybrids.
Advances in Human Genetics. 3, 173-235.

Stern, C. (1936). Somatic crossing over and segregation in *Drosophila
melanagaster*. *Genetics*. 21, 625-730.

Stoker, M.P.G. (1972). Tumour viruses and the sociology of fibroblasts.
Proceedings of the Royal Society B. 181. 1-17.

Van Someren, H., Westerveld, A., Hagemeijer, A., Mees, J.R., Meera Khan,
P., and Zaalberg, O.B. (1974). Human antigen and enzyme markers in
man-Chinese hamster somatic cell hybrids. *Proceedings of the
National Academy of Sciences, U.S.A.* 71, 962-965.

Weiss, M.C., and Green, H. (1967). Human-mouse hybrid cell lines containing partial complements of human chromosomes and functioning
human genes. *Proceedings of the National Academy of Sciences, U.S.A.*
58, 1104-1111.

GENE AMPLIFICATION IN PLANTS

A. SIEGEL

Biology Department,
Wayne State University,
Detroit, Michigan 48202, U.S.A.

The question of gene amplification in higher plants
is reviewed briefly. Evidence is lacking for the
type of massive amplification of genes for ribosomal
RNA which occurs in *Xenopus* and other animals, but
data are available to indicate differential replication of portions of the plant genome during development.

Chloroplast DNA is considered in terms of differential
replication of plant genetic material. Experimental
data are presented which reveal that 4% of tobacco leaf
cell, but only 0.77% of root cell DNA is chloroplast DNA.
Two-thirds of leaf cell chloroplast DNA is found to reside
in the nucleus rather than in chloroplast and this is
true of almost all of root cell chloroplast DNA. The
question of whether the chloroplast DNA found in the
nucleus is part of the chromosomal DNA complement or is
episomal DNA, is discussed.

The term gene amplification has been used to describe the differential replication of a specific region of the genome in particular
cells of an organism (Brown and Dawid, 1968). The term was originally applied to the fairly massive replication in amphibian oocytes
of the genes which code for ribosomal RNA. The magnitude of this
asynchronous replication is such that in *Xenopus laevis* these genes
are increased some 700-fold (Brown and Dawid, 1968; Gall, 1969).
Gene amplification has been found to occur not only in amphibian
oocytes but in those of insects and a range of other animals. There
are a number of examples of differential replication of other portions
of genomes in specific cells and tissues of animals, some of which
will be discussed in other papers of this volume. Among these are the
"under-replication" of heterochromatin during endoduplication of euchromatin in the salivary glands of Drosophila (Hennig and Meer, 1971)
and the origin of mammalian lymphocyte cytoplasmic membrane DNA as an
asynchronously replicated portion of the nuclear genome (Meinke *et al.*,

1973).

There have been a number of reports concerning differential replication of portions of plant genomes and a brief review of these will be given here.

An early report by Chen and Osborne (1970) indicated that wheat embryos have about 50% more ribosomal DNA than do ungerminated wheat embryos (Table 1). Similar studies comparing seed and leaf DNA of two *Nicotiana* species indicate a reverse situation; the seed DNA having about half the content of genes for ribosomal RNA than does leaf DNA (Lightfoot, 1972). However, in not all cases examined does there prove to be an apparent difference between seed and leaf DNA in this regard because none has been found in Chinese cabbage (Table 1).

Table 1

		% DNA hybridized to ribosomal RNA
Wheat[1]	embryo	0.32
	germinated embryo	0.2
Nicotiana tabacum[2]	seed	0.15
	leaf	0.3
Nicotiana paniculata[2]	seed	0.43
	leaf	0.7
Brassica pekinensis[2]	seed	1.7
	leaf	1.8

[1] Chen and Osborne (1970)
[2] Lightfoot (1972)

Several comments are in order concerning the type of data summarized in Table 1. The differences observed are at most two-fold and, as a consequence, special care must be taken to ensure that they are not spurious in nature. As a matter of fact the data for wheat shown in Table 1 could not be reproduced by Ingle and Sinclair (1972) who concluded that there was no change in rDNA content of wheat embryo cells during germination. They, together with others, have pointed to technical details which must be carefully controlled in making comparative estimates of rDNA content such as differential DNA extraction, differential sticking of DNA to nitrocellulose membranes, purity of DNA preparations, etc. Thus, because of the several experimental pitfalls, the rest of the data in Table 1 await independent confirmation.

The massive type of rDNA gene amplification observed by animal workers is so far restricted to a single cell type, the oocyte. In comparing different developmental stages, tissues or organs of a plant, however, different cell types are lumped together. It is possible, therefore, that massive amplification in a single cell or cell type may have gone undetected or observed only marginally in the types of experiments that have been reported. Ingle and Sinclair (1972) made an attempt to overcome this difficulty by estimating rDNA content of tissue from very early developmental stages of plant embryos and were unable to find evidence for gene amplification.

We are left with no strong evidence for massive amplification of the genes for ribosomal RNA in plants and perhaps many plants have solved the problem of an occasional requirement for a large number of copies of these genes by carrying the large number in the genome rather than having a smaller number which is increased when required. A comparison of the number of genes for high-molecular-weight rRNA and 5S RNA in *Xenopus* and pumpkin (*Cucurbita pepo* L.) lends support to this idea. It can be seen in Table 2 that the somatic cells of the frog contain a much greater number of genes for 5S RNA than for high-molecular-weight rRNA; the genes for 5S RNA are not amplified in the oocyte whereas those for the other rRNAs are amplified many fold.

Table 2

Number of gene copies per haploid genome

	HMW rRNA	5S RNA
Cucurbita pepo[1]	3400	3900
Xenopus laevis[2]	450-650	9000-24000

[1]Thornburg and Siegel (1973)
[2]Brown and Weber (1968); Birnstiel *et al.* (1972)

In contrast, in pumpkin, where rDNA amplification has not been observed, the genes for high-molecular-weight rRNA and 5S RNA are approximately equal in number. Furthermore, plants in general contain more copies of the rRNA genes than do animals (Birnstiel *et al.*, 1971) and for these reasons it may be that gross amplification of rDNA may be found not to occur in higher plants.

There have been reports that a rather dense (1.722 g/cc) species of DNA appears in certain plant tissues cultured under stress conditions and this has been attributed to amplification of a portion of the genomes (Quétier et al., 1968; Guille et al., 1968). However, it has been pointed out that this newly appearing density satellite component probably results from contaminating bacterial DNA (Pearson and Ingle, 1972). More recently the transient appearance of a similar dense satellite component correlated with cytological events has been observed in the DNA extracted from *Nicotiana glauca* primary pith explants cultured under aseptic conditions (Parenti et al., 1973). The functional significance of this dense satellite component and whether or not it can be considered an example of gene amplification remain to be elucidated.

Perhaps the most compelling evidence for differential replication of portions of the plant genome comes from the work of Pearson et al. (1974) who observed that the DNA extracted from different tissues of melons have different proportions of a 1.706-1.708 g/cc satellite component and further that the precise density of the satellite (as well as of the main band) is tissue dependent. This phenomenon and its interpretation is presented in greater detail in the paper by Ingle in this volume.

I discuss now the results of experiments which indicate an example of a unique sort of gene amplification (or differential replication) in plants. The uniqueness lies in the consideration of an organelle DNA, - that contained in chloroplasts. The question asked is whether the amount of chloroplast DNA in leaf cells, which are green because they contain chloroplasts, differs from root cells which are not green and if so, by how much? A second type of question asked is whether chloroplast DNA is found in nuclei as well as in chloroplasts and if so, how much? The reason for inquiring into this second question is that there have been several reports that the DNA extracted from chloroplasts and nuclei have at least some nucleotide sequences in common. Among these are those that hybridize to chloroplastic ribosomal RNA (Tewari and Wildman, 1970), certain species of t-RNA (Williams et al., 1973) and tobacco mosaic virus pseudovirion RNA (Siegel, 1971). In addition, it has been demonstrated that chloroplast and nuclear DNA hybridize with each other although the number of matching sequences could not be determined in the type of experiment performed (Kung *et al.*, 1972).

The procedure adopted for this study is that devised by Gelb *et al.*
(1971) in which renaturation kinetics of the DNA in question is observed in the presence and in the absence of an excess of test DNA.
In the present instance, the renaturation of tritium-labelled tobacco
chloroplast DNA was measured in the presence of an excess of either
calf-thymus DNA or plant DNA extracted from leaves, leaf nuclei, roots
or root nuclei. The rate of renaturation of the isotopically labelled
chloroplast DNA is increased by the extent to which the excess unlabelled DNA contains nucleotide sequences homologous to those in the
chloroplast DNA. The extent of renaturation was determined by measuring the amount of trichloroacetic acid precipitable radioactivity remaining after digesting single-stranded DNA with S_1 nuclease.

The data accumulated from several experiments are presented in
Figures 1 and 2. The concentration of DNA in the annealing mixtures
of all of the experiments was 400 µg/ml and contained either 1 or 2
µg of ^{3}H-chloroplast DNA, the rest being calf thymus DNA, test DNA or
mixtures of calf thymus and test DNA.

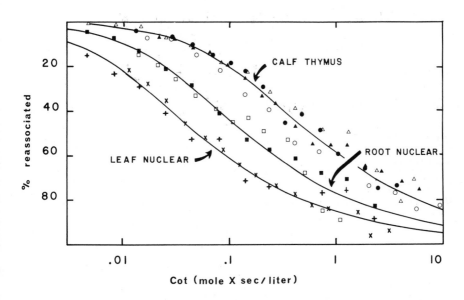

Fig. 1
(See overleaf for legend)

20

Fig. 1. Reassociation of 1 μg/ml ^{3}H-chloroplast DNA (14,000 cpm/μg) in the presence of 400 μg/ml of either calf thymus DNA (o,•, Δ,▲separate experiments), root nuclear DNA (□,■) or leaf nuclear DNA (+,x). The curves are derived from the equation: % reassociated = 100{1-1/1+kCot)$^{0.44}$}*, assuming k to be 6.67, 27 and 76.9 for the calf thymus, root nuclear and leaf nuclear DNA curves, respectively. Mixtures of chloroplast and test DNA were sheared to a single-strand size *ca*. 450 nucleotides with the aid of a VirTis 60 homogenizer operated at 50,000 rpm for 30 mins, precipitated by addition of 2 volumes of ethanol, resuspended in 0.2 M acetate buffer pH 6, passed through a Chelex-100 column, reprecipitated with ethanol and resuspended in an appropriate concentration of sodium phosphate buffer, pH 7 (Cot values in the figure are corrected to 0.12 M buffer) (Britten *et al.*, 1974). Twenty μl portions in sealed capillary tubes were placed in a boiling water bath for 5 mins and subsequently incubated at 60°C for different time periods, the samples were digested at 40°C for 1 hr with S$_1$ nuclease (Miles Lab.) (Ando, 1966) and spotted onto 3 MM filter paper discs. The discs were washed with cold 5% TCA (3x), ethanol-ethyl ether 1:1, ether and radioactivity determined in a Packard scintillation spectrometer using a toluene based fluid containing 0.5 ml NCS (Amersham-Searle).

Nuclei and chloroplasts were obtained by chopping washed young leaves in Honda medium and sedimenting a resuspended 1000 x g pellet of the resultant brei through a discontinuous gradient, containing layers of 60%, 45% and 25% sucrose (Tewari and Wildman, 1966). Chloroplasts were collected from the 45-25% discontinuity and the DNA extracted (Siegel, 1971). Nuclear DNA was obtained from the material sedimented to the bottom of the tube after it had been washed exhaustively with a solution containing 0.5 M sucrose, 0.01 M MgCl$_2$, 0.01 M CaCl$_2$, 0.005 M mercaptoethanol, 0.05 M Tris, 3.5% Triton x-100 (Rohm and Haas), pH 8.2. Root nuclei and nuclear DNA were obtained in a similar manner. Total DNA was obtained from washed leaves or roots by freezing in liquid N$_2$, pounding to a fine powder, adding 4 vol. of a 1:1 mixture of phenol and buffer (0.1 M EDTA, 4% SDS, 0.2 M Tris, pH 8) and allowing the mixture to thaw while mixing. The aqueous phase was collected, heated to 60°C, incubated at 50°C in the presence of ribonuclease for two hrs (50 μg/ml pancreatic, 750 units/ml T$_1$) and overnight with the addition of 100 μg/ml predigested Pronase. The mixture was reextracted with phenol, precipitated with ethanol, resuspended in 5 mM sodium phosphate buffer, pH 7 and passed twice through an A-15 Bio-gel (Bio-Rad) column.

Radioactive chloroplast DNA was prepared by infiltrating 50 gm shredded young leaves with an equal weight of 0.01M KH$_2$PO$_4$ containing one mCi ^{3}H-thymidine and incubating 10-12 hrs under fluorescent lighting. The tissue was then chopped, fractionated and the chloroplast DNA extracted.

---

* This equation has been found to describe the reassociation kinetics of DNA when determined with the aid of S$_1$ nuclease (Smith, M., personal communication).

The rate constant for reassociation of 1 µg/ml sheared (ca. 450 nucleotides) labelled chloroplast DNA in the presence of 400 µg/ml of calf thymus DNA was found to be 6.67 (Fig 1). The validity of the experimental procedure was tested by observing the effect on the rate constant of adding 5 µg/ml of unlabelled chloroplast DNA to the re-annealing mixture. It was found that this 6-fold increase in concentration increased the rate constant 6-fold to 40 (Fig 2) in line with theoretical expectation.

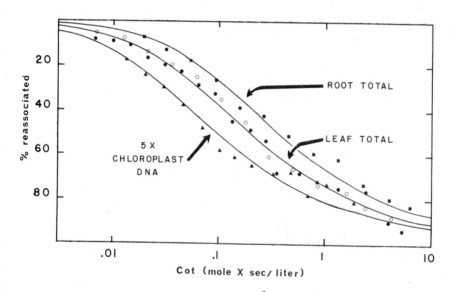

Fig. 2. Reassociation of 1 or 2 µg/ml ^{3}H-chloroplast DNA in the presence of either a 5-fold excess of unlabelled chloroplast DNA, a 50-fold excess of total leaf DNA or a 75-fold excess of total root DNA. The concentration of DNA in the reaction mixtures was made to 400 µg/ml with addition of calf thymus DNA. The curves assume rate constants of 40, 20 and 10.5 for the unlabelled chloroplast DNA, total leaf DNA and total root DNA reactions, respectively. See legend to Figure 1 for experimental details.

The effect on the rate constant of substituting leaf or root nuclear or total DNA for all or part of the calf thymus DNA is listed in
Table 3. The rate constant is increased when nuclear, as well as
total DNA from either leaves or roots is present in the annealing mixture indicating that nuclei as well as chloroplasts contain chloroplast DNA. The calculated number of copies of chloroplast DNA per
nucleus and cell are presented in Table 4. The calculations were made
according to Gelb et al. (1971) assuming a molecular weight for tobacco chloroplast DNA of 9.3×10^7 Daltons (Kolodner and Tewari, 1972)
and a DNA cell content of 6×10^{12} Daltons (Siegel et al., 1973).
These data show, not surprisingly, that leaf cells contain on average
more chloroplast DNA (about 5 times more) than do root cells and provides an example of differential replication of a species of cellular
DNA. The degree of difference, however, is not of the magnitude of
rDNA amplification in animal oocytes. The results have been obtained
with mature tissues of one species and it is not unlikely that either
larger or smaller differences will be found at different developmental
stages or with different species.

Table 3

Rate constants for annealing of ^{3}H-chloroplast
DNA in the presence of unlabelled DNAs.

[3]H chloroplast DNA	Unlabelled DNA	No. of Assays	Rate Constant
1 µg	400 µg calf thymus	4	6.67
1 µg	5 µg chloroplast 400 µg calf thymus	1	40.0
1 µg	400 µg leaf nuclear	2	76.9
2 µg	400 µg leaf nuclear	2	41.7
1 µg	400 µg root nuclear	2	27.0
2 µg	400 µg root nuclear	1	16.7
2 µg	100 µg leaf total 300 µg calf thymus	2	20.0
2 µg	150 µg root total 250 µg calf thymus	1	10.5

Reannealing mixture

Table 4
Number of copies of chloroplast DNA
per nucleus and cell

	no. of copies	% of the cell's DNA
leaf nucleus	1695	2.6
leaf cell	2580	4.0
root nucleus	488	0.76
root cell	499	0.77

In contrast to genomic (chromosomal) DNA, it is not unexpected to find different amounts of an organelle DNA in different cells or tissues of a eucaryote. In this case, however, the organelle DNA is apparently also found in the nucleus. The data indicate that about two-thirds of leaf cell and almost all of root cell chloroplast DNA is found in nuclei rather than in chloroplasts. The first question that comes to mind concerning this observation is whether it is real or perhaps due to some experimental artifact, an obvious one being the possible contamination of nuclear DNA with DNA from chloroplasts during the fractionation and isolation of the organelles. I do not think this likely for a number of reasons:

1. Independently prepared preparations of nuclear DNA yield the same result.
2. Leaf nuclei when obtained by sedimentation of a re-suspended 1000 x g pellet of a leaf brei (prepared by razor blade chopping in Honda medium) through a discontinuous sucrose gradient contain very little green material. This is subsequently removed by rigorous washing in Triton x-100 solutions before DNA extraction.
3. It is difficult to imagine that the fractionation procedure employed would distribute two-thirds of DNA contained in chloroplasts to the nuclear fraction.

If we conclude that the major portion of chloroplast DNA is indeed located in the nucleus rather than in chloroplasts, then we may inquire as to whether at least part of it is chromosomal DNA as ordinarily understood or, alternatively, may be sequestered in the nucleus as an extrachromosomal (episomal?) component. Evidence that at least some chloroplast DNA nucleotide sequences are actually physically attached to chromosomal DNA is provided by the results of the joint reannealing studies of Kung *et al.* (1972) with chloroplast and nuclear DNAs of broad bean and I have obtained a similar result with tobacco DNAs. If chloroplast DNA actually proves to be a constituent of the chromosomal DNA complement then a dilemma is presented in attempting to account for the several examples of cytoplasmic inheritance that are attributed, with good evidence, to chloroplast DNA (i.e. Wong-Staal and Wildman, 1973).

In summary, evidence for differential replication of the genome of higher plants is reviewed briefly. New data are presented which demonstrate that leaf cells contain more chloroplast DNA than do root cells. This may constitute another example of amplification of a portion of the chromosomal genome because it is likely that much of a cell's chloroplast DNA may actually be located in the nucleus rather than in plastids.

ACKNOWLEDGEMENTS

Part of the work reported herein was supported by NSF grant GB 38666X and AEC contract AT(11-1)-2384. The excellent technical assistance of Ms. Kathryn Kolacz is gratefully acknowledged.

REFERENCES

Ando, T. (1966). A nuclease specific for host-denatured DNA isolated from a product of *Aspergillus oryzae*. *Biochimica et Biophysica Acta.* 114, 158-168.

Birnstiel, M., Chipchase, M., and Spiers, J. (1971). The ribosomal RNA cistrons. *Progress in Nucleic Acid Research and Molecular Biology.* 11, 351-389.

Birnstiel, M., Sells, B., and Purdom, I. (1972). Kinetic complexity of RNA molecules. *Journal of Molecular Biology.* 63, 21-39.

Britten, R., Graham, D., and Neufeld, B. (1974). Analysis of repeating DNA sequence by reassociation. *Methods in Enzymology.* 29, 363-418.

Brown, D., and Dawid, I. (1968). Specific gene amplification in oocytes. *Science.* 160, 272-280.

Brown, D., and Weber, C. (1968). Gene linkage by RNA-DNA hybridization. I. Unique DNA sequences homologous to 4S RNA, 5S RNA and ribosomal RNA. *Journal of Molecular Biology.* 34, 661-680.

Chen, D., and Osborne, D. (1970). Ribosomal genes and DNA replication in germinating wheat embryos. *Nature.* 225, 336-340.

Gall, J. (1969). The genes for ribosomal RNA during oogenesis. *Genetics.* 61 (supplement), 121-132.

Gelb, L.D., Kohne, D.E., and Martin, M.A. (1971). Quantitation of
   Simian Virus 40 sequences in African green monkey, mouse and
   virus transformed cell genomes. *Journal of Molecular Biology*.
   57, 129-145.

Guille, E., Quetier, F., and Huquet, T. (1968). Etude des acides
   desoxyribonucléiques des végétaux. Formation d'un ADN nucléaire
   riche en G + C lors de la blessure de certaines plantes supérieures.
   Comptes rendus de l'Academie des Sciences. Serie D. 266, 836-838.

Hennig, W., and Meer, B. (1971). Reduced polyteny of ribosomal RNA
   cistrons in giant chromosomes of *Drosophila* hydei. *Nature New
   Biology*. 233, 70-72.

Ingle, J., and Sinclair, J. (1972). Ribosomal RNA genes and plant
   development. *Nature*. 235. 30-32.

Kolodner, R., and Tewari, K. (1972). Molecular size and conformation
   of chloroplast deoxyribonucleic acid from pea leaves. *Proceedings
   of the National Academy of Sciences, U.S.A*. 247, 6355-6364.

Kung, S., Moscarello, M., and Williams, J. (1972). Studies with chloroplast and mitochondrial DNA. *Biophysics Journal*. 12, 474-483.

Lightfoot, D. (1972). Quantitation of ribosomal genes in developing
   Chinese cabbage and tobacco plants. Ph.D. Dissertation, University
   of Arizona.

Meinke, W., Hall, M., Goldstein, D., Kohne, D.E., and Lerner, R. (1973).
   Physical properties of cytoplasmic membrane-associated DNA. *Journal
   of Molecular Biology*. 78, 43-56.

Parenti, R., Guille, E., Grisvard, J., Durante, M., Giorgi, L., and
   Buiatti, M. (1973). Transient DNA satellite in dedifferentiating
   pith tissue. *Nature New Biology*. 246, 237-239.

Pearson, C., and Ingle, J. (1972). The origin of stress-induced
   satellite DNA in plant tissues. *Cell Differentiation*. 1, 43-51.

Pearson, G., Timmis, J., and Ingle, J. (1974). The differential replication of DNA during plant development. *Chromosoma*. 45, 281-294.

Quetier, F., Guille, E., and Vedel, F. (1968). Etude des acides
   desoxyribonucleiques des végétaux. Isolement et propriétés d'un
   ADN nucléaire riche en G + C. *Comptes rendus de l'Academie des
   Sciences, Serie D*. 226, 735-738.

Siegel, A. (1971). Pseudovirions of tobacco mosaic virus. *Virology*.
   46, 50-59.

Siegel, A., Lightfoot, D., Ward, O., and Keener, S. (1973). DNA
   complementary to ribosomal RNA: relation between genomic proportion and ploidy. *Science*. 179, 682-683.

26

Tewari, K., and Wildman, S.G. (1966). Chloroplast DNA from tobacco leaves. *Science*. 153, 1260-1271.

Tewari, K., and Wildman, S.G. (1970). The information content in chloroplast DNA. *Symposium of the Society of Experimental Biology*. 24, 147-179.

Thornburg, W., and Siegel, A. (1973). Characterization of the rapidly reassociating deoxyribonucleic acid of *Cucurbita pepo* L. and the sequences complementary to ribosomal and transfer ribonucleic acids. *Biochemistry*. 12, 2759-2765.

Williams, G., Williams, A., and George, S. (1973). Hybridization of leucyl-transfer-ribonucleic acid isoacceptors from green leaves with nuclear and chloroplast deoxyribonucleic acid. *Proceedings of the National Academy of Sciences, U.S.A.* 70, 3498-3501.

Wong-Staal, F., and Wildman, S.G. (1973). Identification of a mutation in chloroplast DNA correlated with formation of defective chloroplasts in a variegated mutant of *Nicotiana tabacum*. *Planta*. 113, 313-326.

ENVIRONMENTALLY INDUCED DNA CHANGES IN FLAX

C. A. CULLIS

John Innes Institute,
Colney Lane,
Norwich.

An environmentally induced DNA change in the flax
variety Stormont Cirrus has been investigated. An
induced line, L, which had a high DNA, and a low
DNA line, S, showed no gross differences from the
original variety, Pl, when analysed in neutral
caesium chloride gradients. Reassociation experiments showed that DNA from L contained a class of
moderately repeated sequences which was not present
in the DNA from either S or Pl. L was also shown
to have a lower proportion of highly repeated
sequences than either S or Pl. S and Pl showed the
same renaturation kinetics. The number of ribosomal
cistrons was determined and it was found that L and
Pl had the same number of cistrons while S had 70%
fewer cistrons than L and Pl.

Certain environmental conditions can induce genetic changes in
particular varieties of plants (Durrant, 1962; Perkins *et al.*, 1971).
Under some conditions these changes can be stable and inherited over
many generations (Durrant, 1971). One of the best examples of this
environmental conditioning occurs in the flax variety Stormont Cirrus.
When this variety is grown for the first five weeks in a heated
greenhouse, in particular combinations of fertiliser and soil pH, it
can be changed from its original form (Pl) to either a large stable
genotroph, L, or a small stable genotroph, S, depending on the ferti-
lisers supplied. L plants can be up to six times the weight of S
plants depending on the environment in which they are grown, with Pl
plants having a weight between those of L and S (Durrant, 1962). Both
L and S are shorter than Pl, with L being taller than S.

L and S behave as two distinct genetic types giving equilinear
inheritance when they are reciprocally crossed and no transmission
through reciprocal grafts (Durrant, 1962). L can have up to sixteen
per cent more nuclear DNA than S as shown by Feulgen staining while
Pl has a DNA content between L and S, having ten per cent less DNA

than L, but six per cent more DNA than S (Evans, 1968).

DNA was extracted from the three types L, S and Pl and analysed to characterise the DNA differences between the types. The DNA can be resolved by neutral caesium chloride equilibrium centrifugation into a main band, density 1.698 g.cm^{-3}, a less dense satellite, density 1.688 g.cm^{-3} comprising about fifteen per cent of the total DNA, and heavy shoulder, density 1.709 g.cm^{-3} (Cullis, 1973; Ingle et al., 1973). Analyses of the DNA prepared from L, S and Pl plants showed no significant differences in the distribution between these three gross fractions.

The DNA from all three genotrophs has been further studied by following the renaturation of purified, denatured DNA. The renaturation of DNA can be followed in two ways: (i) The separation by hydroxyapatite chromatography of single- and double-stranded material, formed by incubation for different times. (ii) By following the reduction in absorbance at 260 nm as double-stranded regions are formed (for review of methods see Britten et al., 1974).

HYDROXYAPATITE FRACTIONATION.

The rate of renaturation and the proportion of unique and repeated sequences as measured by hydroxyapatite chromatography depends on the size of the DNA used in the reaction (Davidson et al., 1973). $C_o t$ curves for S DNA of different fragment size are shown in Figure 1. The size of fragments is an approximate weight-average size and is not intended to suggest a homogeneous sample. As can be seen from Figure 1, at all fragment sizes used the $C_o t$ curve is biphasic, suggesting the presence of both repeated and unique sequences in the flax genome. The proportion of repeated sequences in the genome found by this method depends on the fragment size of the DNA. However if the double-stranded DNA separated from the hydroxyapatite column was thermally denatured and the hyperchromicity determined, then an estimate of the extent of the double-stranded regions could be obtained. From the values in Table 1 it can be seen that in this way a consistent value for the proportion of the genome made up of highly repeated sequences in S was obtained.

$C_o t$ curves obtained with DNA extracted from plants of the Pl variety were indistinguishable from those obtained with DNA from S.

$C_o t$ curves obtained with DNA from L were triphasic rather than iphasic (Fig 2). The curve for L DNA could be separated into three

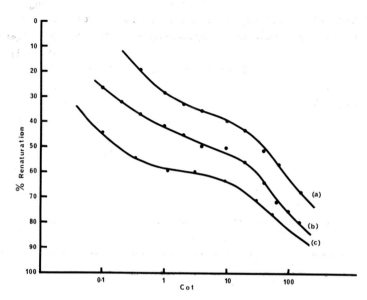

Fig. 1. C_0t curves of S DNA sheared to different fragment size. DNA was sheared in a Dawe sonicator to different sizes in 0.12M phosphate buffer (PB) (equimolar sodium dihydrogen phosphate and disodium hydrogen phosphate). The molecular weight was determined by band sedimentation (Marmur, 1961). The sheared DNA was incubated in 0.12M PB at 60° and then diluted to 0.03M PB and passed over an hydroxyapatite column also maintained at 60°C. The single stranded DNA was eluted with 0.12M PB at 60°C and the double stranded DNA was eluted with 0.4M PB. Zero time binding was determined as described by Davidson *et al.* (1973), and subtracted from all values. (a) DNA sheared to average length of 600 nucleotides. (b) DNA sheared to average length of 1400 nucleotides. (c) DNA sheared to average length of 1900 nucleotides.

parts which are called fast, intermediate and slow. The fast portion of the curve included the DNA renaturing up to a C_0t of 0.5; the intermediate and slow fractions had some overlap with the intermediate renaturing from C_0t 0.5 to C_0t 20 and the slow fraction from C_0t 10 to C_0t 1000.

The fast renaturing fraction of the DNA from L renatures at a similar rate to the faster fraction in S and Pl, but comprises a smaller proportion of the total DNA in L than in S and Pl (Fig 3 and Table 1). This is true at all fragment lengths tested (Table 1). In S and Pl the faster renaturing fraction of the DNA is twenty-nine per cent of the total while in L it is only twenty-four per cent of the total DNA.

TABLE 1

Proportion of highly repeated DNA in L and S with different
fragment size as measured by hydroxyapatite chromatography.

Fragment size (nucleotides)	% highly repeated DNA		Corrected % highly repeated DNA (1)	
	L	S	L	S
2,000	53	60	23.4	29.4
1,400	36	43	24.2	28.5
600	31	35	25.0	28.0

(1) Corrected for reduced hyperchromicity due to unpaired
lengths.

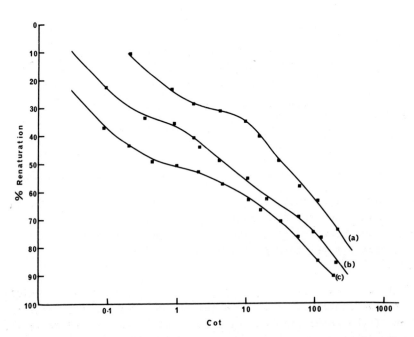

Fig. 2. $C_o t$ curves of L DNA sheared to different fragment size. Other
conditions as for Figure 1.

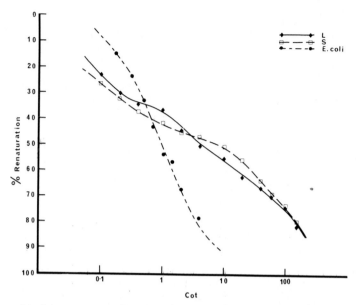

Fig. 3. C₀t curves of L, S and *E.coli* DNA sheared to an average
length of 1400 nucleotides, incubated and fractionated as
described for Figure 1.

The difference between L and S and Pl in the slow renaturing
fraction is difficult to estimate due to the overlap between the intermediate and slow fractions in L and the variation in amount of fast
and slow fractions with differing fragment size.

The intermediate fraction was discernible at all three fragment
sizes at which the DNA from L has been renatured (Fig 2). This
suggested that the fraction was not closely associated with the fast
renaturing fraction of the DNA (Davidson *et al.*, 1973). At the
higher fragment sizes the proportion of DNA renaturing in the intermediate fraction was higher than that which could be explained by the
increased DNA. If this fraction was isolated and thermally denatured
and the percentage double-stranded material determined, then consistent results were obtained, with the intermediate fraction comprising thirteen per cent of the total DNA of L. It is particularly
relevant that this amount would account for all the extra DNA in the
L type.

OPTICAL RENATURATION.

The renaturation of DNA from L, S and Pl was also followed optically and the results plotted as the reciprocal of the remaining
hyperchromicity against time (Wetmur and Davison, 1968). This type
of plot is shown in Figure 4 with *E. coli* included as a reference DNA.

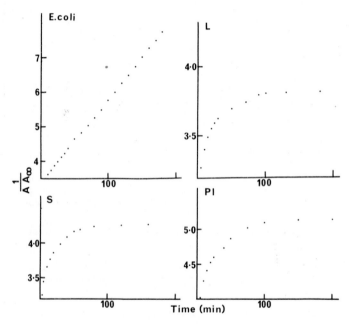

Fig. 4.   DNA sheared to 1200 nucleotides was denatured in 3.5 x SSC
(SSC-0.15M sodium chloride, 0.01M trisodium citrate) 40%
formamide and renatured at 40°C.   Absorbance was continuously monitored in a Unicam SP1700 spectrophotometer.   The
reciprocal of the remaining hyperchromicity was plotted
against time.

For a DNA composed entirely of unique sequences a straight line should
be obtained (Wetmur and Davison, 1968) as is observed with *E. coli*.
For the flax DNAs this is not the case and the results show a heterogeneity within the repeated sequences. It is possible to estimate
the proportion of fast renaturing sequences from this type of plot
and the values obtained are twenty-six per cent for L, thirty per
cent for S and Pl. These values are in close agreement with those
obtained in Table 1.

THE NATURE OF THE DNA DIFFERENCES BETWEEN L, S AND Pl.

The difference in the fast renaturing fraction between L and S
can be explained by the dilution of this fraction in L, if it is
assumed that none of the extra DNA in L occurs in the fast renaturing
fraction (Cullis, 1973). The amount of DNA occurring in the intermediate fraction in L would account for the total difference between
L and S. Because of its position in the $C_o t$ curve, this intermediate
fraction could only have been derived by a large increase in part of
the slow renaturing portion of the DNA. By comparison with the rate
of reassociation of *E. coli* and from the known nuclear DNA amounts
in flax, it can be shown that the slow renaturing fraction of the DNA
must be composed of unique sequences. By a similar comparison of the
reassociation rates of the intermediate and slow fractions, the intermediate sequences must be present at a frequency about seventy-times
that of the unique sequence.

S has six per cent less DNA than Pl but the reassociation of the
DNAs from these two types is indistinguishable. For this to be the
case the reduction in S must have occurred in both the fast and slow
renaturing sequences in approximately the same proportions as these
sequences are present in the genome. Knowing the differences in total
DNA between L and Pl, the differences between the fast renaturing
fractions of L and Pl cannot be explained on a simple dilution model.
It could be explained if L had lost the same sequences as S when being induced from Pl. In this case both L and S differ from Pl in that
they have lost a particular set of sequences, while L further differs
from both by having gained a large number of copies of another set of
sequences. To conclusively demonstrate this situation it would be
necessary to isolate the set of sequences by which L and S differ from
Pl but this has not yet been achieved.

RIBOSOMAL CISTRON DIFFERENCES.

The ribosomal RNA (r-RNA) genes are highly repeated in all eukaryotes in which they have been studied. Between plant species there is
wide variation in the number of ribosomal RNA genes (Ingle and
Sinclair, 1972) and also variation within a plant species (Phillips
et al., 1973).

The number of r-RNA genes in the flax lines was determined by DNA-
RNA hybridisation. Purified 25S and 18S r-RNA was hybridised to DNA
extracted from leaves of each of the three types, L, S and Pl. The
results are given in Table 2. It can be seen that the number of r-RNA

genes in L and Pl are approximately the same but are some seventy per cent higher than the number in S. Thus the induction of the S type has resulted in a large decrease in the number of r-RNA genes present although the induction of the L genotroph has had no effect on the number of r-RNA genes. It has also been shown that the density in caesium chloride of the r-RNA cistrons is higher in S than it is in L (Timmis and Ingle, 1973).

TABLE 2

DNA content and number of r-RNA genes
in L, S and Pl genotrophs

		L	S	P1	%L/S
DNA (10^{-12}g per 2C nucleus)		1.53	1.32	1.40	1.16
% Hybridisation	(i)	0.514	0.355	0.629	
	(ii)	0.429	0.280	0.419	
	(iii)	0.390	0.312	0.405	
No. of genes/2C nucleus	(i)	2632	1568	2947	1.68
	(ii)	3294	1855	2944	1.78
	(iii)	4492	3106	4269	1.45

(i) Hybridised with pea 1.3 plus 0.7 x 10^6 r-RNA (4 + 2μg ml^{-1}).
(ii) Hybridised with pea 1.3 x 10^6 r-RNA (4μg ml^{-1}). (iii) Hybridised with pea 0.7 x 10^6 r-RNA (4μg ml^{-1}).

It has recently been shown (Pearson et al., 1974) that differential replication of satellite DNA can occur during plant development, so that different tissues contain a different proportion of the satellite DNA. It has also been shown that in Drosophila the number of ribosomal cistrons can be increased or decreased depending on the chromosomal constitution (Ritossa, 1972). Thus mechanisms appear to exist in both plants and animals for a differential amplification of various portions of the genome. In the flax variety Stormont Cirrus these mechanisms appear to be affected by environmental factors to such an extent that large changes in the DNA can occur. The results show that many kinds of alterations can be involved in the induction of L and S from Pl. These can involve an increase in a particular part of the unique sequences in the case of L, a decrease at a particular locus (the ribosomal cistrons) in S, or a more generalised loss in

both repeated and unique sequences in the case of L and S. Thus the
plasticity of the Pl variety appears to exist at many loci and the
total extent to which the genome can be modified may not yet have
been observed.

REFERENCES

Britten, R.J., and Kohne, D.E. (1968). Repeated sequences in DNA.
Science. 161, 528-540.

Britten, R.J., Graham, D.E., and Neufeld, B.R. (1974). Analysis of
repeating DNA sequences by reassociation. *Methods in Enzymology*.
29, 363-407.

Cullis, C.A. (1973). DNA differences between flax genotrophs. *Nature*.
243, 515-516.

Davidson, E.H., Hough, B.R., Amenson, C.S., and Britten, R.J. (1973).
General interspersion of repetitive with non-repetitive sequence
elements in the DNA of *Xenopus*. *Journal of Molecular Biology*.
77, 1-23.

Durrant, A. (1962). The environmental induction of a heritable change
in *Linum*. *Heredity*. 17, 27-61.

Durrant, A. (1971). The induction and growth of flax genotrophs.
Heredity. 27, 277-298.

Evans, G.M. (1968). Nuclear changes in flax. *Heredity*. 23, 25-38.

Ingle, J., Pearson, G.G., and Sinclair, J. (1973). Species distribution and properties of nuclear satellite DNA in higher plants.
Nature New Biology. 242, 193-197.

Ingle, J., and Sinclair, J. (1972). Ribosomal RNA genes and plant
development. *Nature*. 235, 30-32.

Marmur, J. (1961). A procedure for the isolation of deoxyribonucleic
acid from microorganisms. *Journal of Molecular Biology*. 3, 208-
218.

Pearson, G.G., Timmis, J.N., and Ingle, J. (1974). The differential
replication of DNA during plant development. *Chromosoma*. 45, 281-
294.

Perkins, J.M., Eglington, E., and Jinks, J.L. (1971). The nature of
the inheritance of permanently induced changes in *Nicotiana
rustica*. *Heredity*. 27, 441-457.

Phillips, R.L., Wang, S.S., Weber, D.F., and Kleese, R.A. (1973).
The nucleolar organiser region (NOR) of maize: a summary.
Genetics. 74, Supplement 2, 212.

Ritossa, F. (1972). Procedure for magnification of lethal deletions

36

for genes of ribosomal RNA. *Nature New Biology.* 240, 109-111.

Timmis, J.N., and Ingle, J. (1973). Environmentally induced changes
   in r-RNA gene redundancy. *Nature New Biology.* 244, 235-236.

Wetmur, J.G., and Davison, N. (1968). Kinetics of renaturation of DNA.
   Journal of Molecular Biology. 31, 349-370.

A ROLE FOR DIFFERENTIAL REPLICATION OF DNA IN DEVELOPMENT

J. INGLE AND J.N. TIMMIS

*Department of Botany, University of Edinburgh,
Mayfield Road, Edinburgh. EH9 3JH.*

A brief summary is given of the experimental evidence which
demonstrates that the eucaryote genome does not necessarily remain constant during cell development, but varies
as the result of differential replication of the DNA.
The replicative state of the DNA is considered in two
particular types of nuclear development, polytene nuclei
in *Phaseolus* suspensor cells and polyploid nuclei in
melon and cucumber fruit. Development of *Phaseolus* polytene chromosomes appears to involve complete genome replication, which contrasts with the development of polyteny in *Drosophila*. The production of fruit polyploid
nuclei, however, involves differential replication of
the genome on the basis of both heterochromatin-euchromatin proportions and quantitative and qualitative
satellite DNA analysis.

The usual starting point in studies on cellular differentiation is
the assumption that all the cell types of an individual have the same
genome, or multiples thereof, and that differentiation results from
the selective use of certain parts of this genome. Expression of the
genome may be regulated at the level of RNA transcription, translation into protein or at post-translational levels. How valid is this
assumption of genome constancy? The constancy of the genome within
the germ line of an individual is certainly inherent in the study of
genetics, and the phenomenon of totipotency of plant cells is frequently interpreted as indicating that all cells contain the full, or
constant genome. However, it is far from clear exactly how many, and
what types of cell exhibit totipotency. Given that the first stage
in the expression of totipotency is cell division, all those cells
capable of division may be considered as potentially totipotent, and
hence contain the full genome. Even this assumption, although valid
qualitatively, need not hold quantitatively for those DNA sequences
present in the genome as multiple copies. It is probable therefore,
that the germ lines and the dividing cells within an individual maintain a constant genome, that is the complete genome is fully replica-

ted prior to division, but there is little reason for assuming such constancy during the differentiation of the various cell types which make up the bulk of the plant.

In this paper we would like to briefly summarise the experimental evidence which demonstrates that the genome does not necessarily remain constant, but varies as the result of differential replication of the DNA, and then consider in more detail our own studies on the differentiation of polyploid cells during fruit development.

THE DEMONSTRATION OF DIFFERENTIAL REPLICATION

The data available in the literature at present are rather limited, due to the poor development of DNA fractionation. The progress in such technology at the present time promises, however, that much better data should soon be available.

DNA PER NUCLEUS.

Many tissues contain cells with 2c, 4c, 8c and higher ploidy DNA contents. If the genome is fully replicated at each stage, then the nuclear DNA values determined by microspectrophotometry of Feulgen stained material will fall within a doubling series. Although such doubling series are apparent in some tissues (Fox, 1969), other tissues show excellent examples of non-doubling series (Fox, 1970) (Table 1). These non-doubling DNA distributions indicate that the genome is under-replicated at each synthetic phase, and furthermore the distributions appear to be tissue specific.

Table 1

Doubling and non-doubling series in insect tissues

Mean DNA values expressed relative
to the 2c value

	2c	4c	8c	16c	32c
Dermestes haemorrhoidalis					
Testis wall	2.00	4.00	8.06	15.96	32.12
Fat body	2.00	3.72	8.22	16.86	-
Schistocerca gregaria					
Midgut diverticulum	2.00	3.44	4.96	8.78	-
Malpighian tubule	2.00	3.32	5.68	9.32	13.14
Fat body	2.00	4.06	6.80	11.56	20.44

Data taken from Fox (1969, 1970)

HETEROCHROMATIN - EUCHROMATIN DISTRIBUTION.

Fractionation of the genome on the basis of heterochromatin-euchromatin content shows that differential replication of these two components may occur within a tissue. In the majority of nuclei in the testis wall of *Dermestes maculatus* the replication of heterochromatin and euchromatin remain in step, but a small population of nuclei exists in which the heterochromatin is under-replicated by 2 or 3 rounds compared with the euchromatin (Fox, 1971). An over-replication of heterochromatin has been described in cultured orchid protocorms, which contain two populations of nuclei, one the result of regular endopolyploidy and the other containing a much larger content of heterochromatin (Nagl, 1972). Cucumber fruit tissue also contains two populations of polyploid nuclei, one with under-replication and the other with over-replication of the heterochromatin (Pearson *et al.*, 1974) (see Fig 4, d and e).

DNA *FRACTIONATION.*

DNA may be fractionated on the basis of sequence repetition. In certain cases the repetitious fraction differs from the bulk of the DNA in buoyant density and is resolved as a minor component on analytical CsCl equilibrium centrifugation. Any differential replication of this particular DNA fraction is therefore easily observed in Model E analytical studies, and this approach will be considered in more detail later in the paper.

RIBOSOMAL-RNA GENE REDUNDANCY.

The replicative status of at least one specific DNA sequence, that coding for ribosomal RNA (r-RNA), may be readily and accurately determined by hybridisation of high specific activity r-RNA back to denatured DNA. In fact the classic example of differential replication is that of the r-RNA genes during oocyte development in *Xenopus laevis.* In the somatic cells of Xenopus about 0.3% of the genome is complementary to r-RNA, but this is amplified several thousand fold in the oocyte, such that it finally represents 70% of the nuclear DNA. Gene amplification of this magnitude appears to be restricted to the Amphibia, although 10 fold amplification of the r-RNA genes has been observed in the cricket (Lima-de-Faria *et al.*, 1969), and water beetle (Gall *et al.*, 1969). Hybridisation studies with DNA from plant tissues indicate that no comparable amplification of the r-RNA genes occurs during normal plant development (Ingle and Sinclair, 1972).

In addition to these gross changes in r-RNA gene redundancy, re-

ferred to as amplification, much smaller variations occur, a phenomenon which will be called gene flexibility to distinguish it from
the previous examples. Perhaps the best example of this occurs in
Drosophila, where the r-RNA genes, located at the nucleolar organising
regions (NOR) are on the sex chromosomes. In a normal individual
each NOR contains 250 copies, but when the X chromosome is present by
itself, or with another sex chromosome containing a NOR deletion, the
gene redundancy increases to approximately 400. When the chromosome
is transferred back into an individual containing another X or Y, the
redundancy decreases to 250 (Tartof, 1971). Similarly the "bobbed"
NOR (a mutation of the NOR in which the number of r-RNA genes is reduced) rapidly accumulates genes when maintained within a "bobbed"
nucleus, but the number is reduced when introduced into a wild-type
nucleus (Henderson and Ritossa, 1970). The degree of reiteration of
the gene is therefore controlled by other factors within the nucleus.
Such a control of r-RNA gene redundancy is also indicated from studies
on euploids and aneuploids of hyacinth. The number of genes per NOR
was constant in the 2x, 3x and 4x euploids, but was lower in all the
aneuploids examined (Table 2), possibly due to the chromosomal inbalance of these aneuploid nuclei (Timmis *et al.*, 1972). A similar
situation exists in wheat, where the aneuploids are genetically
better characterised (Mohan and Flavell, 1974). Addition or removal
of the 1B nucleolus organizing regions increases or decreases the
r-RNA gene redundancy respectively, as would be expected simply on a
dosage basis (Table 3). Although removal of the 1A NORs greatly reduces the redundancy, addition of two extra 1A chromosomes does not
result in the expected increase. Addition of 5D chromosomes actually
decreases the overall redundancy, whereas removal of the 5D NOR results in no decrease. These results with the 5D chromosomes suggest
the presence of factors associated with this chromosome which depress
the overall r-RNA gene redundancy. Intraspecific flexibility of the
r-RNA gene is also shown in the case of environmentally induced genotrophs of flax, where r-RNA gene redundancy is 70% greater in the
large compared with the small genotroph, although the difference in
total nuclear DNA content is only 16% (Timmis and Ingle, 1973). These
examples demonstrate that the multiplicity of r-RNA genes, and by
extrapolation possibly other genes, is flexible in plants, the level
being controlled by other DNA sequences within the genome.

Table 2

Ribosomal RNA genes in hyacinth

Chromosome complement	NOR number	Genes 2c nucleus	Genes/NOR
Euploids			
2x	2	17,284	8,642
3x	3	22,575	7,525
4x	4	31,910	7,978
		Mean	8,048
Aneuploids			
3x - 1	2	12,690	6,345
2x + 1	3	19,626	6,542
2x + 3	3	22,461	7,487
4x - 2	3	19,935	6,645
4x - 3	4	26,350	6,588
		Mean	6,721

Data taken from Timmis *et al.* (1972)

Table 3

Ribosomal RNA genes in *Triticum aestivum*

(Chinese Spring)

Nucleolar organisers				r-RNA genes (% of euploid control)
1B	1A	5D	6B	
2	2	2	2	100
4	2	2	2	128
0	2	2	2	84
2	4	2	2	101
2	0	2	2	58
2	2	4	2	76
2	2	0	2	105

Data taken from Mohan and Flavell (1974)

DIFFERENTIATION OF POLYTENE CELLS.

All four of the experimental approaches outlined above have been applied to the study of differentiation of salivary gland cells in *Drosophila*. The popularity of this tissue is of course due to its suitability for genetical and cytological studies, and to the relative ease of isolation of these cells, and hence comparatively pure polytene chromosome DNA. Spectrophotometric measurements of Feulgen-stained nuclei show that during the development of the polytene chromosomes the total absorbance falls below multiples of that present in diploid nuclei. This appears to be due to a reduction in the amount of chromocentric heterochromatin (Rudkin, 1965). A comparison of the renaturation kinetics of DNA prepared from polytene and non-polytene cells shows that the polytene genome contains only 5% of fast renaturing, or repetitious DNA, compared with 20% in the non-polytene cells (Dickson *et al.*, 1971). This heterochromatic repetitious DNA is at least in part analogous to satellite DNA, since Gall *et al.* (1971) showed by Model E analysis that satellite DNA is greatly under-replicated in the polytene chromosomes of both *Drosophila melanogaster* and *D. virilis*. The r-RNA genes are also under-replicated during this development, the number of copies being only 40 compared with 280 per haploid genome present in a diploid cell (Hennig and Meer, 1971).

The development of polytene chromosomes in the suspensor cells of *Phaseolus coccineus* has similarly been studied. Feulgen microspectrophotometry indicates however, a full replication of the DNA up to 8192c (Brady, 1973), and there is no indication of under-replication of the satellite DNA (Fig 1). A similar percentage of satellite is present in suspensor and embryo cells. The greater percentage of satellite DNA present in these cells (34-35%) compared with 24% in leaf tissue (Ingle *et al.*, 1973) will be discussed later in relation to the differential replication of satellite DNA in melon tissues. There is no information yet regarding the replicative state of the r-RNA genes. The development of polyteny in *Phaseolus*, and probably also in mouse, where normal replication of satellite DNA is observed in the foetal trophoblast cells (Sherman *et al.*, 1972), is therefore very different from the extensive differential replication of the *Drosophila* polytene genome.

DIFFERENTIATION OF FRUIT TISSUE.

In initial studies on the presence of satellite DNA in different plant species (Ingle *et al.*, 1973) it was established that satellite,

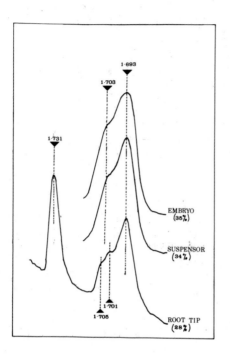

Fig 1   CsCl Model E analysis of *Phaseolus coccineus* tissues.

Root tips, dissected suspensors and embryos were disrupted in
1% sodium dodecyl sulphate, 50 mM tris-HCl pH 8.0 and 10 mM EDTA,
then made up with CsCl to an initial density of 1.707 g.cm^{-3} and
centrifuged to equilibrium (20 hrs) at 44000 rpm at 25°C in a
Beckman Model E analytical centrifuge. *Micrococcus lysodeikticus*
DNA was included in each sample as an internal marker at
1.731 g.cm^{-3}. (Figures in brackets refer to % satellite DNA).

when present in a species, was present in all the tissues of that
species. However, as more data was accumulated for melon (*Cucumis
melo*) it became clear that the amount of the satellite DNA, expressed
as a percentage of the total, varies in different tissues (Fig 2).
The values fall into two classes, a high percentage of satellite in
seed and root tip and a lower percentage in leaf, cotyledons and fruit.
Certain tissues, such as flower and hypocotyl appear to be intermediate in value (Table 4). The variation in the percentage satellite in
Phaseolus tissues is consistent with this picture. The errors involved in the preparation of DNA, the Model E fractionation and the
calculation of the percentage satellite DNA and buoyant densities of

44

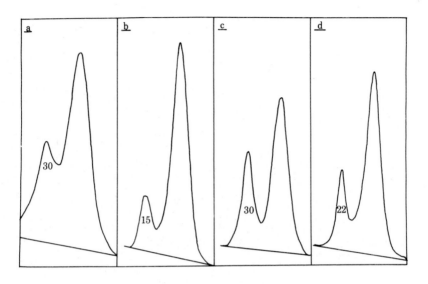

Fig 2   Analyses of DNA prepared from melon tissues.

Total DNA was prepared from the seeds (a) and the fruit (b) of
a Spanish Winter melon. Seeds of this melon were germinated and
DNA prepared from root tips (c) and hypocotyl tissue (d). The
DNAs were analysed by Model E centrifugation. The value indicates
the amount of satellite as a percentage of the total DNA. Taken
from Pearson *et al.* (1974).

---

mainband and satellite peaks have been assessed by preparation and
Model E analysis of 6 replicate samples of melon seed DNA.   The
satellite component, expressed as a percentage of the total DNA,
varies from 23.3 to 27.0, with a mean of 25.0 ± 1.5%.   The buoyant
densities of the mainband and satellite are 1.6922 ± 0.0002 and
1.7057 ± 0.0002 g. cm^{-3} respectively. Furthermore the percentage
satellite DNA does not vary significantly with different DNA preparative procedures, nor with the size of the DNA being fractionated
(Pearson *et al.*, 1974).   The difference in percentage satellite in
the various tissues of melon, for example 26.1% in Hero of Lockinge
seed and 14.4% in cotyledon tissue (Table 4), is far greater than the
experimental error involved in preparation and analysis, which is of
the order of a 3% spread.

The fruit and seeds of an individual melon show the largest difference in the percentage of satellite DNA, and since commercially

Table 4

Variation in the percentage satellite DNA
in melon tissues

Melon variety	Satellite DNA (% of total DNA)						
	Seed	Root tip	Flower	Hypoco- tyl	Leaf	Cotyle- don	Fruit
Hero of Lockinge	26.1	25.8	21.5	19.7	15.6	14.4	17.1
Honeydew	29.4	-	-	22.4	-	17.2	-
Musk	29.5	-	-	-	18.9	17.6	17.1
Spanish winter	27.7	-	-	21.9	-	17.9	18.5
"	24.3	-	-	-	16.2	19.2	16.7
"	30.3	29.7	-	21.9	-	20.4	15.2
"	24.8	-	-	-	16.3	14.1	20.9
"	34.1	-	-	-	-	20.9	23.9
"	36.3	-	-	-	15.1	19.3	-

Data taken from Pearson *et al.* (1974).

obtained melons are a convenient source of material, further investigation was concentrated on these tissues. With all of the melons
examined the fruit tissue contains a smaller percentage of satellite
DNA than the corresponding seeds (Table 5). The reason for the variation in the amount of satellite between individuals of the same
variety of melon e.g. Honeydew or Spanish Winter, is not clear. It
is probably due to inadequate identification and genetic variation.
The classification of melons is very confused since it is a highly
polymorphic species with hybridisation readily occurring between the
different forms. The variental names, with the exception of Hero of
Lockinge, refer to types rather than individual cultivars. With the
range of material analysed the difference between percentage satellite
in fruit and seed is highly significant (p < 0.1%). Not only is the
amount of satellite different in these tissues, the buoyant density
of the satellite is also significantly different (p < 0.1%). This
difference in buoyant density may indicate that different satellites
are present in these two tissues, different perhaps with respect to
methylation of the bases. However, satellite DNA in melon contains
at least two types of sequence with very different properties (Sinclair *et al.*, 1974) so it is possible that fruit and seed satellites
contain different proportions of these components. There is also a
small, but highly significant (p < 0.1%) difference in the fruit and

Table 5

Percentage satellite and buoyant densities of mainband and satellite DNAs from seed and fruit

Variety	Satellite DNA (% of total)		Buoyant density (g. cm^{-3})			
			Satellite		Mainband	
	Seed	Fruit	Seed	Fruit	Seed	Fruit
Melon						
Hero of Lockinge	26.1	17.1	1.7058	1.7085	1.6924	1.6938
Honeydew	29.2	21.8	1.7062	1.7076	1.6919	1.6933
"	24.7	18.4	1.7056	1.7079	1.6922	1.6936
"	23.4	18.1	1.7056	1.7066	1.6922	1.6926
"	25.5	17.3	1.7063	1.7082	1.6928	1.6939
Ogen	30.6	18.2	1.7056	1.7088	1.6927	1.6937
"	24.9	19.1	1.7063	1.7067	1.6929	1.6930
Musk	24.2	16.3	1.7057	1.7070	1.6923	1.6927
"	29.5	17.1	1.7056	1.7071	1.6926	1.6940
Spanish winter	27.7	18.5	1.7063	1.7075	1.6917	1.6931
"	24.3	16.7	1.7068	1.7076	1.6931	1.6932
"	30.3	15.2	1.7064	1.7079	1.6923	1.6928
"	24.8	20.9	1.7057	1.7070	1.6918	1.6928
"	34.1	23.9	1.7056	1.7069	1.6921	1.6925
Cucumber						
Kariha[*]	41	34	I 1.7011	1.7034	1.6939	1.6943
			II 1.7060	1.7084		

[*] Data are the means of 7 samples. The two satellite components are combined for the percentage satellite DNA determination.

Data taken from Pearson et al. (1974).

---

seed mainband DNA buoyant density. A similar situation exists with cucumber tissues, where again percentage satellite, and the buoyant densities of the two satellites differ between fruit and seed (Table 5). These differences in the percentage satellite and in the buoyant densities of both satellite and mainband DNAs indicate massive differential replication. The r-RNA genes, however, appear to be normally replicated in these tissues since the percentage hybridisation of the mainband DNA (75 and 82% of the total DNA in seed and fruit respectively) is constant (Table 6).

Table 6

r-RNA gene redundancy in melon tissues

	Seed	Fruit
% satellite DNA	25	18
% hybridisation 1.3×10^6 rRNA (2 μg/ml, 3h)	0.19	0.22
$1.3 + 0.7 \times 10^6$ rRNAs (2+1 μg/ml, 3h)	0.29	0.32
1.3×10^6 rRNA (5 μg/ml, 1 hr)	0.16	0.20

Data taken from Ingle *et al.* (1974).

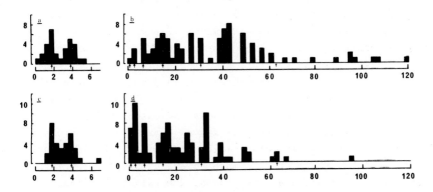

Fig 3 The nuclear DNA content of cucumber and melon tissues.
Nuclei were isolated from root tip (a) and fruit tissue (b) of
cucumber and root tip (c) and fruit tissue (d) of melon, and
the DNA content was determined by Feulgen microspectrophotometry.
The arrows indicate the doubling series based on the 2c and 4c
values of the root nuclei. Taken from Pearson *et al.* (1974).

The bulk of the tissue in melon and cucumber seeds is the young
cotyledonary tissue, which from Feulgen microspectrophotometry is very
similar to root tips, containing only 2c and 4c nuclei (Fig 3a and c).
The nuclear complement of cucumber fruit tissue is very different,
with few 2c and 4c nuclei, the majority being between 16c and 64c
(Fig 3b). The melon fruit similarly contains a high proportion of
polyploid nuclei (Fig 3d). Although the melon fruit nuclear distri-

bution contains peaks corresponding to the doubling series, many nuclei are intermediate, particularly between the 16c, 32c and 64c values. Cucumber fruit shows a more continuous nuclear distribution, the bulk of the nuclei being outside of the doubling series, particularly above 16c. One possible explanation is that even in the mature fruit many nuclei are still in S phase. However, in cucumber fruit the continuous distribution is in part the result of the presence of at least two classes of nuclei differing in heterochromatin content (Fig 4, compare d and e). Measurements made on root tip nuclei indicate approximately 25% heterochromatin (Fig 4c), whereas the majority of fruit nuclei contain only 5% heterochromatin (Fig 4d). From a comparison with the heterochromatin-euchromatin distribution in root tip 2c and 4c nuclei it appears that these fruit nuclei have 16c to 64c euchromatin but only 2c to 8c heterochromatin. A minority of the fruit nuclei contain 37% heterochromatin (Fig 4e), and appear to have undergone an additional round of heterochromatin replication. This analysis shows that during fruit development there is both under- and over- replication of the heterochromatin in different nuclei (Fig 5), the overall change being a reduction of heterochromatin. A similar under-replication of heterochromatin is indicated during the development of melon fruit nuclei (Fig 4a and b) but the heterochromatin areas are too diffuse for quantitative analysis. The differentiation of fruit cells is therefore accompanied by, or perhaps the result of, differential replication of the genome, demonstrated in these studies both in terms of chromatin and DNA fractionation.

Fig 4 Nuclei isolated from cucumber and melon tissues.

Feulgen stained nuclei from melon root tip, with 4c DNA content (a); melon fruit, 32c, (b); cucumber root tip, 4c, (c); cucumber fruit, euchromatin 16c, heterochromatin 4c, (d); cucumber fruit, euchromatin 16c, heterochromatin 32c, (e). Taken from Pearson *et al*. (1974).

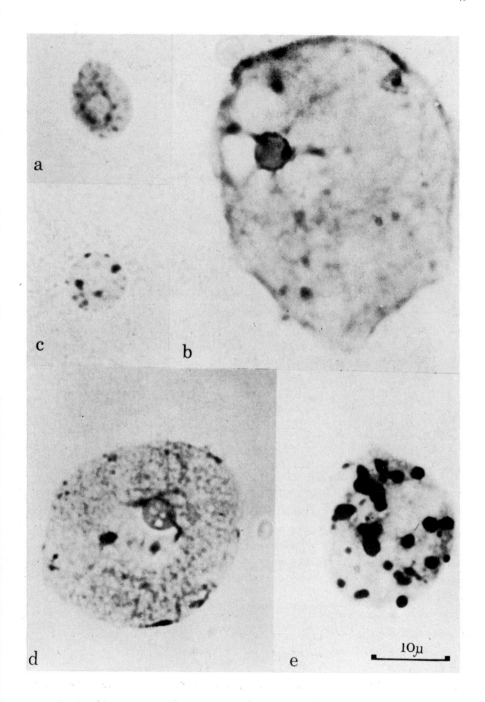

a

b

c

d

e

10μ

50

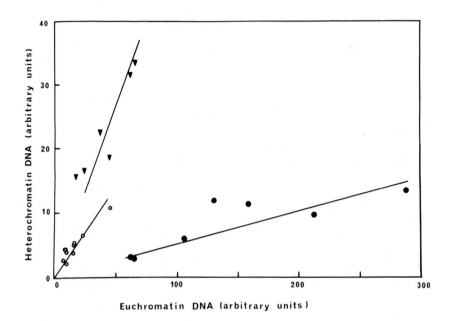

Fig 5 Heterochromatin-euchromatin content of cucumber nuclei.

Nuclei were isolated from root tips and fruit, stained with
Feulgen and photographed. The heterochromatin and euchromatin
was estimated from the area of heterochromatin and the average
difference in density between euchromatin and heterochromatin,
determined from microdensitometer scans of photographs of 9
root tips and 12 fruit nuclei. Root tip nuclei contained 25%
heterochromatin (———o———), and fruit nuclei were of two classes
containing 37% (———▼———) and 5% (——— ● ———) heterochromatin.
Taken from Pearson et al. (1974).

The type of results presented in this paper indicate that the eucaryote genome is dynamic rather than static, since the flexibility
of r-RNA gene redundancy demonstrates a mechanism capable of both
sensing and reacting to the nuclear environment. The results also
show that differential replication may play a major role in cellular
differentiation.

REFERENCES

Brady, T. (1973). Feulgen cytophotometric determination of the DNA
    content of the embryo proper and suspensor cells of *Phaseolus
    coccineus*. *Cell Differentiation*. 2, 65-75.

Dickson, E., Boyd, J.B., and Laird, C.D. (1971). Sequence diversity
    of polytene chromosome DNA from *Drosophila hydei*. *Journal of*

Molecular Biology. 61, 615-627.

Fox, D.P. (1969). DNA values in somatic tissues of Dermestes. I Abdominal fat body and testis wall of the adult. *Chromosoma*. 28, 445-456.

Fox, D.P. (1970). A non doubling DNA series in somatic tissues of the locusts *Schistocerca gregaria* and *Locusta migratoria*. *Chromosoma*. 29, 446-461.

Fox, D.P. (1971). The replicative status of heterochromatic and euchromatic DNA in two somatic tissues of *Dermestes maculatus*. *Chromosoma*. 33, 183-195.

Gall, J.G., Cohen, E.H., and Polan, M.L. (1971). Repetitive DNA sequences in *Drosophila*. *Chromosoma*. 33, 319-344.

Gall, J.G., MacGregor, H.C., and Kidston, M.E. (1969). Gene amplification in the oocytes of *Dytiscid* water beetles. *Chromosoma*. 26, 169-187.

Henderson, A., and Ritossa, F.M. (1970). On the inheritance of rDNA of magnified bobbed loci in *D. melanogaster*. *Genetics*. 66, 463-473.

Hennig, W., and Meer, B. (1971). Reduced polyteny of r-RNA cistrons in giant chromosomes of *Drosophila hydei*. *Nature*. 233, 70-72.

Ingle, J., Pearson, G.G., and Sinclair, J. (1973). Species distribution and properties of nuclear satellite DNA in higher plants. *Nature New Biology*. 242, 193-197.

Ingle, J., and Sinclair, J. (1972). Ribosomal RNA genes in plant development. *Nature*. 235, 30-32.

Ingle, J., Timmis, J.N., and Sinclair, J. (1974). The relationship between satellite DNA, ribosomal-RNA gene redundancy and genome size. *Plant Physiology* (In press).

Lima-de-Faria, A., Birnstiel, M., and Jaworska, H. (1969). Amplification of ribosomal cistrons in the heterochromatin of *Acheta*. *Genetics Supplement*. 61, 145-159.

Mohan, J., and Flavell, R.B. (1974). Ribosomal RNA cistron multiplicity and nucleolar organisers in hexaploid wheat. *Genetics*. 76, 33-44.

Nagl, W. (1972). Evidence of DNA amplification in the orchid *Cymbidium in vitro*. *Cytobios*. 5, 145-154.

Pearson, G.G., Timmis, J.N., and Ingle, J. (1974). The differential replication of DNA during plant development. *Chromosoma*. 45, 281-294.

Rudkin, G.T. (1965). Non-replication of DNA in giant chromosomes. *Genetics*. 52, 470 (Abstract).

Sherman, M.I., McLaren, A., and Walker, P.M.B. (1972). Mechanism of accumulation of DNA in giant cells of mouse trophoblast. *Nature New Biology*. 238, 175-176.

Sinclair, J., Wells, R., and Ingle, J. (1974). The complexity of satellite DNA in a higher plant. *Biochemical Journal* (In press).

Tartof, K.D. (1971). Increasing the multiplicity of ribosomal RNA genes in *Drosophila melanogaster*. *Science*. 171, 294-297.

Timmis, J.N. and Ingle, J. (1973). Environmentally induced changes in r-RNA gene redundancy. *Nature New Biology*. 244, 235-236.

Timmis, J.N., Sinclair, J., and Ingle, J. (1972). Ribosomal RNA genes in euploids and aneuploids of hyacinth. *Cell Differentiation*. 1, 335-339.

QUANTITATIVE VARIATION IN NUCLEOLAR RIBOSOMAL RNA GENE MULTIPLICITY IN WHEAT AND RYE

R. B. FLAVELL

Cytogenetics Department,
Plant Breeding Institute,
Cambridge, CB2 2LQ. England.

The number of ribosomal RNA genes is not constant in *Triticum aestivum* and *Secale cereale*. A variation greater than two fold has been found among different plant varieties. Using substitution lines in which each of the nucleolar organisers of the wheat variety Chinese Spring has been replaced in turn, by homologues from other wheat varieties, genetic variation in ribosomal RNA gene number has been detected on each of the four nucleolar organiser chromosomes of wheat. In lines in which a major nucleolar organiser has been deleted, some evidence has been gained for a compensating amplification, presumably at one or more of the other nucleolar organisers. Evidence is presented that some rye supernumerary B chromosomes carry ribosomal RNA genes.

The variation in rRNA gene multiplicity provides evidence for the existence of a genetically controlled mechanism for monitoring and altering rRNA gene multiplicity in higher plants.

DNA sequences easily quantitatively studied are those specifying the ribosomal RNAs. This is because the gene products, the ribosomal rRNA molecules, can be purified and used in rRNA/DNA hybridisation experiments to determine the proportion of the DNA in the genome that is complementary to rRNA. The genes specifying the two major rRNA molecules are clustered, in almost all eucaryotes so far investigated, to form the chromosomal sites where nucleoli are organised. Within each cluster or nucleolar organiser, the two different rRNA sequences are probably universally alternately repeated but separated by 'spacer' sequences (Birnstiel *et al.*, 1971; Ingle, 1973).

It has been known for many years that plants contain many thousands of rRNA genes and that different species contain different numbers of rRNA genes and nucleolar organisers (Birnstiel *et al.*, 1971; Matsuda and Siegel, 1967; Goldberg *et al.*, 1972; Ingle and Sinclair, 1972).

However, only relatively recently has it been confirmed that in higher plants, for example hyacinth (Timmis *et al.*, 1972), flax (Timmis and Ingle, 1973), maize (Phillips *et al.*, 1973) and wheat (Mohan and Flavell, 1974; Flavell and Smith, 1974a, b), variation in rRNA gene multiplicity can occur within a species. Over the past two years I and three colleagues, Mr. Derek Smith, Dr. Jag Mohan and Dr. Jurgen Rimpau, have been investigating the genetic control of rRNA gene multiplicity in wheat and rye. The results presented in this paper have been chosen to illustrate the kind of variation which occurs in rRNA gene multiplicity in these species and the chromosomal location of this variation. These cereal species are particularly suitable for such a study because of the availability of an impressive array of specially constructed genotypes, based on one genetic background, carrying nucleolar organiser chromosomes in different doses or from different sources (Sears, 1954; Law, 1968; Law and Worland, 1972).

In all experiments DNA was highly purified from green plant tissue by methods described in detail elsewhere (Smith and Flavell, 1974; Mohan and Flavell, 1974). Equilibrium banding in caesium chloride gradients (Flamm *et al.*, 1966) was included in the purification procedure. ^{3}H labelled rRNA was highly purified from wheat embryos after germination in ^{3}H uridine for 20 hrs at 25°C (Mohan and Flavell, 1974). Ribosomal RNA/DNA hybridisation was carried out by incubating approximately 20 μg of denatured DNA, immobilised on nitrocellulose filters, with 5 μg/ml rRNA at 70°C in 6 x SSC (SSC = 0.15 M sodium chloride 0.015 M sodium citrate) for 3 hrs. The proportion of the genome that is complementary to rRNA was calculated in each case from the amount of ribonuclease-resistant rRNA bound to the filters and the amount of DNA on each filter (Mohan and Flavell, 1974; Flavell and Smith, 1974a, b).

NUCLEOLAR ORGANISER VARIATION IN rRNA *GENE MULTIPLICITY IN WHEAT AND RYE.*

The percentages of the DNAs of seven hexaploid wheat and two rye varieties that hybridise to rRNA are shown in Table 1. There is approximately a 2.2 fold variation in the number of rRNA genes within the array of wheat varieties. All the values are significantly different from one another except Hope, Koga II and Cappelle-Desprez. Approximately 2.5 times more rRNA hybridised to Petkus rye DNA than to DNA from King II rye.

There are at least four different chromosomes in hexaploid wheat,

Table 1
Hybridisation of rRNA to DNA from different
wheat and rye varieties.

	Variety	% hybridisation (mean[a] ± standard error of mean)
A Wheat	Chinese Spring	.086 ± .004
	Hope	.072 ± .004
	Koga II	.073 ± .002
	Cheyenne	.111 ± .005
	Cappelle-Desprez	.075 ± .003
	Triticum spelta	.120 ± .004
	Holdfast	.055 ± .001
B Rye	Petkus	.174 ± .002
	King II	.071 ± .002

[a] mean of 6 replicate filters
Data from Flavell and Smith, 1974a

viz. 1A, 1B, 5D and 6B able to form nucleoli when isolated in micro-
nuclei at meiosis (Crosby, 1957). In order to test whether the observed variation in rRNA gene multiplicity in some of the wheat varieties
(Table 1), could be ascribed to individual nucleolar organisers, the
rDNA content was investigated in substitution lines derived from the
variety Chinese Spring. In these lines each of the nucleolar organiser
chromosomes of Chinese Spring viz. 1A, 1B, 5D and 6B has been replaced
in turn by its homologue from the hexaploid varieties Hope, Cappelle-
Desprez and *Triticum spelta*. The Chinese Spring/Hope substitution
lines were constructed by Dr. E.R. Sears of the University of Missouri,
and the Chinese Spring/*Triticum spelta* and Chinese Spring/Cappelle-
Desprez substitution lines by Dr. C.N. Law and his colleagues in this
Department at the Plant Breeding Institute, Cambridge. All these substitution lines have had between 5 and 7 backcrosses to Chinese Spring
during the course of their construction, so it is reasonable to assume
that the lines differ from Chinese Spring only in the substituted
chromosome. The percentages of the DNAs from these lines which hybridise to rRNA are shown in Table 2. Replacement of Chinese Spring
chromosome 1A by chromosome 1A from Cappelle-Desprez or from *Triticum
spelta* resulted in an increase in the number of rRNA genes. Replacement of Chinese Spring chromosome 6B by its homologue from Cappelle-
Desprez, Hope or *Triticum spelta* resulted in a significant reduction

in rRNA gene multiplicity. Chromosome 1B from *Triticum spelta* and chromosomes 5D from *Triticum spelta* and Hope also caused significant differences in rRNA hybridisation.

Since at least one example within each group of lines in which the same nucleolar organiser chromosome has been replaced, was significantly different from Chinese Spring, it can be concluded that variation in rRNA gene multiplicity exists in wheat at each of the known nucleolar organiser sites. Furthermore, the sum of the effects of each of the nucleolar organiser chromosomes from each variety, when substituted into Chinese Spring, can account for the differences between the rRNA gene contents of Chinese Spring and the donor varieties. This implies that the effects of the other 17 chromosomes in hexaploid wheat on rRNA gene multiplicity are relatively small or non-existent. These findings, that the rRNA gene content of homologous nucleolar organisers is not constant in different wheat varieties, support similar conclusions drawn from studies on aneuploid stocks of Chinese Spring carrying different doses of each of the nucleolar organiser chromosomes (Mohan and Flavell, 1974).

Table 2

Hybridisation of rRNA to DNAs from Chinese Spring lines
with different substituted nucleolar organiser chromosomes.

Varietal source of substituted chromosome

Substituted Chromosome	Cappelle-Desprez	*Triticum spelta*	Hope
1A	.100 ± .003*	.115 ± .002***	.079 ± .002
1B	.086 ± .002	.077 ± .002*	.089 ± .002***
5D	.087 ± .003	.096 ± .002**	.066 ± .002***
6B	.044 ± .001***	.069 ± .002***	.065 ± .003***

Values are mean % ± standard deviation of mean DNA hybridised to rRNA.

* significantly different from internal Chinese
Spring control at 5% level of probability

** significantly different from internal Chinese
Spring control at 1% level of probability

*** significantly different from internal Chinese
Spring control at .1% level of probability

Data from Flavell and Smith, 1974a

EVIDENCE FOR rRNA *GENE AMPLIFICATION TO COMPENSATE FOR A* rRNA *GENE DEFICIENCY?*

The short arm of chromosome 1B in wheat carries a major nucleolar organiser. An analysis (Flavell and Smith, 1974b) of the rRNA gene content of a range of aneuploid stocks of Chinese Spring, originally isolated by Dr. E.R. Sears (Sears, 1954) has shown that when the number of 1B chromosomes is increased from two to four in tetrasomic 1B lines, the percentage DNA which hybridises to rRNA increases by 31 to 49% (Table 3). However, when the number of 1B nucleolar organisers is reduced from 2 to 0, in the compensated nullisomic 1B or ditelosomic $1B^L$ lines the proportion of the DNA which hybridises to rRNA is reduced by only 15 to 23%. This suggests, but by no means proves, that when a large number of rRNA genes is lost from a major nucleolar organiser, more rRNA genes accumulate elsewhere, presumably at other nucleolar organisers, to partially compensate for the deficiency. Further genetic analyses are in progress to test this hypothesis.

Table 3

The effect of chromosome 1B on rRNA gene content

Genotype	% hybridisation[*]	% of euploid
Euploid	0.090 ± .001	100
Nullisomic 1B Tetrasomic 1A	0.070 ± .001	78
Nullisomic 1B Tetrasomic 1D	0.069 ± .002	77
Ditelosomic $1B^L$	0.076 ± .001	84
Tetrasomic 1B	0.134 ± .002	149
Nullisomic 1A Tetrasomic 1B	0.118 ± .002	131
Nullisomic 1D Tetrasomic 1B	0.128 ± .003	142

[*] Mean ± standard deviation of the mean
Data from Flavell and Smith, 1974b

THE EFFECT OF B CHROMOSOMES OF RYE ON rRNA *GENE CONTENT.*

In the course of studies into the types of repeated nucleotide sequences present in rye supernumerary B chromosomes (Rimpau and Flavell, 1974), we investigated the effect of different numbers of rye B chromosomes on rRNA gene multiplicity. B chromosomes from two different sources were investigated. B chromosomes originally found in a spring rye (*Secale cereale*) from Transbaikal, Siberia were transferred into a Nepal hexaploid wheat variety by Lindström in Sweden (Muntzing, 1973).

From this stock, they were transferred to the wheat variety Chinese Spring by Mr. V. Chapman of this Department by a procedure involving backcrossing to Chinese Spring six times. From the history of crosses from the original rye plant we believe that all the rye B chromosomes in our Chinese Spring stocks should be identical. Hybridisation results using DNAs isolated from plants with 0, 2, 3 and 4 chromosomes are shown in Figure 1. The numbers of genes were calculated from the percentage hybridisation results assuming that each B chromosome increases the DNA per cell by 2.9% (calculated from Jones and Rees, 1968). The results provide good evidence that each of these particular B chromosomes carries approximately 700 rRNA genes in its complement of repeated sequences.

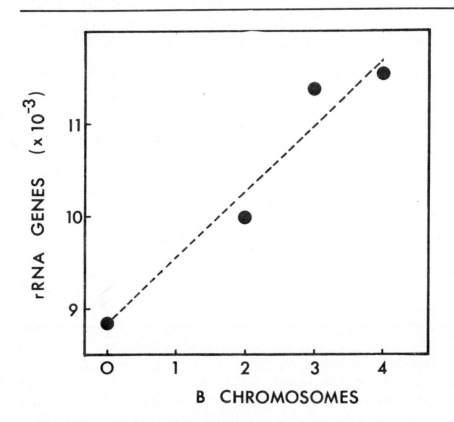

Fig. 1. Effect of rye B chromosomes in wheat on rRNA gene multiplicity. Each point is the mean of six replicate hybridisations. The standard error on each point is approximately ± 300 genes. The B chromosomes came from a *Secale cereale* accession from Transbaikal, Siberia.

The other rye B chromosomes investigated came from an accession of *Secale vavilovi*. This was crossed to a *Secale cereale* line and a population of plants has been maintained from the F_1 for many years in the Department of Agricultural Botany, University of Wales at Aberystwyth. Some seed from this population was kindly supplied to us by Dr. R.N. Jones of that Department. DNA isolated from individual plants carrying 0, 2, 3, or 4 chromosomes was hybridised to rRNA and the number of rRNA genes calculated from the percentage of DNA complementary to rRNA, assuming that each rye B chromosome contained 5.5% of the DNA of the diploid genome (Jones and Rees, 1968).

The results shown in Figure 2 do not provide clear evidence that these rye B chromosomes also carry rRNA genes. There appears to be variation in rRNA gene multiplicity in the population independent of the number of B chromosomes. The linear regression, relating rRNA gene multiplicity and B chromosome number has a positive slope but is not significant (correlation coefficient = .422; P = .056).

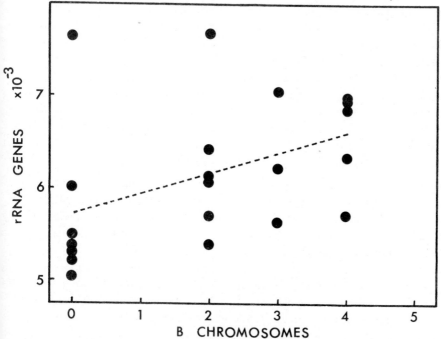

Fig. 2. Effect of rye B chromosomes on rRNA gene multiplicity. Each point is the mean of six replicate hybridisations using DNA isolated from a single plant. The dotted line is the linear regression fitted to the single plant values. (Correlation coefficient 0.422). The B chromosomes come from a *Secale vavilovi* accession.

Whether or not the *S. vavilovi* B chromosomes have rRNA genes, the
B chromosomes from the Transbaikal *S. cereale* accession have many more
rRNA genes than those from the *S. vavilovi* accession. Thus rRNA gene
multiplicity is variable not only on the A chromosomes of rye but also
on the supernumerary B chromosomes.

CONCLUSIONS

The number of rRNA genes in wheat and rye is clearly not constant.
None of the nucleolar organisers in wheat or the nucleolar organisers
of rye A or B chromosomes possesses an invariant number of rRNA genes.
How does this variation in gene multiplicity come about? One can only
conclude that there must be a mechanism in these species for deleting
and amplifying rDNA (Mohan and Flavell, 1974). Similar conclusions
apply to the rRNA gene variation in hyacinth (Timmis *et al.*, 1972),
maize (Phillips *et al.*, 1973), Vicia (Maher and Fox, 1973) and Nicotiana (Siegel *et al.*, 1973).

For any particular genotype, the rRNA gene content seems stable.
Thus the rRNA gene variation does not appear to result from random
uncontrolled changes. This, together with the evidence for compensating rDNA amplification following rDNA deletion (Table 3) and rDNA
variation following polyploidisation or changes in chromosome number
(Siegel *et al.*, 1973; Timmis *et al.*, 1972; Mohan and Flavell, 1974),
suggests that a genetically controlled mechanism exists in some higher
plant species to monitor and regulate rRNA gene redundancy. Genetic
variation in the control of such a mechanism could explain the varietal
variation in nucleolar organiser rRNA gene multiplicity.

In none of the examples of rDNA variation in plants can it be concluded that the variation is unique to rDNA. Neighbouring non rDNA
or unlinked sequences could also be amplified and deleted co-ordinately
with the rDNA. The same mechanism that regulates rDNA deletion and
amplification could also control other sequences elsewhere in the genome. Unless such sequences (1) represent relatively large proportions
of the genome, as do those shown by Pearson *et al.* (1974) to be differ-
entially replicated in plant development or (2) have easily purified
RNA products, it will be more difficult to discover if the mechanism
controlling rRNA gene multiplicity also controls other sequences.

ACKNOWLEDGEMENTS

I am grateful to Mr. Derek Smith, Mr. Michael O'Dell, Dr. Jag Mohan
and Dr. Jurgen Rimpau for their collaboration in the work reported in

this contribution. Dr. Rimpau is supported by the Deutsche Forschungs-
gemeinschaft.

REFERENCES

Birnstiel, M.L., Chipchase, M., and Spiers, J. (1971). The ribosomal
   RNA cistrons. *Progress in Nucleic Acid and Molecular Biology.*
   11, 351-389.

Crosby, A.R. (1957). Nucleolar activity of lagging chromosomes in wheat.
   American Journal of Botany. 44, 813-822.

Flamm, W.G., Bond, H.E., and Burr, H.E. (1966). Density gradient centrifugation of DNA in a fixed angle rotor. *Biochimica et Biophysica
   Acta.* 129, 310-319.

Flavell, R.B., and Smith, D.B. (1974a). Variation in nucleolar organiser
   rRNA gene multiplicity in wheat and rye. *Chromosoma.* 47, 327-334.

Flavell, R.B., and Smith, D.B. (1974b). The role of homologous group
   1 chromosomes in the control of rRNA genes in wheat. *Biochemical
   Genetics.* 12, 271-279.

Goldberg, R.B., Bemis, W.P., and Siegel, A. (1972). Nucleic acid hybridisation studies with the genus *Cucurbita. Genetics.* 72, 253-266.

Ingle, J. (1973). The regulation of ribosomal RNA synthesis. *Annual
   Proceedings of the Phytochemical Society.* 9, 69-91.

Ingle J., and Sinclair, J. (1972). Ribosomal RNA genes and plant development. *Nature.* 235, 30-32.

Jones, R.N., and Rees, H. (1968). The influence of B chromosomes upon
   the nuclear phenotype in rye. *Chromosoma.* 24, 158-168.

Law, C.N. (1968). The development and use of intervarietal chromosome
   substitution. *European Wheat Aneuploid Cooperative Newsletter.*
   1, 22-23.

Law, C.N., and Worland, A.J. (1972). Aneuploidy in wheat and its uses
   in genetic analysis. *Report of Plant Breeding Institute, Cambridge.*
   25-65.

Maher, B.P., and Fox, D.P. (1973). Multiplicity of ribosomal RNA genes
   in *Vicia* species with different nuclear DNA. *Nature* (London)
   New Biology. 245, 170-172.

Matsuda, K., and Siegel, A. (1967). Hybridisation of plant ribosomal
   RNA to DNA; the isolation of a DNA component rich in ribosomal
   RNA cistrons. *Proceedings of the National Academy of Sciences,
   U.S.A.* 58, 673-680.

Mohan, J., and Flavell, R.B. (1974). Ribosomal RNA cistron multiplicity
   and nucleolar organisers in hexaploid wheat. *Genetics.* 76, 33-44.

62

Muntzing, A. (1973). Effects of accessory chromosomes of rye in the
    gene environment of hexaploid wheat. *Hereditas.* 74, 41-56.
Pearson, G.G., Timmis, J.N., and Ingle, J. (1974). The differential
    replication of DNA during plant development. *Chromosoma.* 45, 281-
    294.
Phillips, R.L., Wang, S.S., Weber, D.F., and Kleese, R.A. (1973). The
    nucleolar organiser in maize. *Genetics.* 74, S 212.
Rimpau, J., and Flavell, R.B. (1974). Characterisation of the repeated
    sequence DNA in B chromosomes of rye. (Manuscript in preparation).
Sears, E.R. (1954). The aneuploids of common wheat. *Missouri Agricultu-*
    ral Experimental Station Research Bulletin. 572, 1-59.
Siegel, A., Lightfoot, D., Ward, O.G., and Keener, S. (1973). DNA
    complementary to ribosomal RNA: relation between genomic proportion
    and ploidy. *Science.* 179, 682-683.
Smith, D.B., and Flavell, R.B. (1974). The relatedness and evolution
    of repeated nucleotide sequences in the genomes of some Gramineae
    species. *Biochemical Genetics.* (In press).
Timmis, J.N., and Ingle, J. (1973). Environmentally induced changes
    in rRNA gene redundancy. *Nature New Biology.* 244, 235-236.
Timmis, J.N., Sinclair, J., and Ingle, J. (1972). Ribosomal RNA genes
    in euploids and aneuploids of hyacinth. *Cell Differentiation.* 1,
    335-339.

INTRODUCTORY REMARKS

HERBERT STERN

*University of California,
San Diego.*

I welcome you to our evening session on Transformation. I suspect
that Professor Markham decided to schedule this topic for an evening
rather than the conventional morning or afternoon period in order to
leave us with something to dream about. The speakers may, of course,
upset the plan and inform us that we need not dream any more. Indeed,
to some extent at least, Professor Ledoux has already rendered dream-
ing superfluous, for over the past several years he has published extensively and elaborately on successful genetic transformations by
exogenous application of heterologous DNA to various plant species.
Against such a background of sparkling success, I can at most justify
my participation in this Symposium by drawing attention to the sweep
of mechanisms that must be accommodated to achieve genetic transformation in higher organisms as a consequence of exposing cells to purified DNA preparations. To begin with, there is the need for an efficient uptake of DNA. By itself this is not a simple process because
cells house hydrolytic enzymes which effectively destroy absorbed DNA,
and uptake, however efficient, has little potential for transformation
if accompanied by destruction. Moreover, passage of DNA through the
walls of plant cells is usually difficult; in most plant tissues DNA
may be dissolved by alkali or detergent without being released into
the medium unless the cell walls are broken mechanically or removed
enzymatically. I don't know how extensively pinocytotic mechanisms
operate in plant cells, but if they do function in uptake, they have
not yet been observed to do so across cellulosic walls. Besides,
pinocytotic vesicles are specialized in hydrolyzing their trapped
contents and not in preserving them. A barrier to DNA transformation
thus appears to exist already at the level of DNA uptake and, obviously, there must be an explanation as to how Dr. Ledoux and possibly
others have been able to administer DNA so as to overcome this barrier. However, practical success is not necessarily tied to theoretical insight and there may be a considerable lag between the achieve-
ment and the explanation.

The barrier to genetic transformation by exogenous DNA is not
overcome once the DNA is taken up and retained as such by a cell. The
absorbed DNA must replicate in coordination with its endogenous coun-

terpart in order to be transmitted to cell progeny. We know that in both eukaryotes and prokaryotes there is a lot of metabolic organization governing the initiation and termination of DNA replication. Without such organization synchrony of reproduction among the different chromosomes of a genome and between the genome and the cell as a whole would not exist. Replication of a particular moiety of DNA would probably lead to cell destruction, a situation similar to that produced by infective agents. By contrast, a relatively low rate of replication leads to a loss of the DNA moiety in question and thus to the cell's potential for genetic change. Regardless of how it is effected, coordinated replication and transmission between host and absorbed DNA are clearly essential to heritable transformations.

In addition to its conservation and transmission, the foreign DNA must be transcribed in order to reveal its genetic presence. That, too, is a formidable challenge which Dr. Ledoux appears to have overcome. The complexity of this phase of the transformation process is therefore worth considering. Regulation of gene transcription in higher organisms is a much debated but still poorly understood phenomenon. One source of insight into the phenomenon might be the organization of the high molecular weight RNA molecule found in the so-called HN-RNA fraction. This long molecule is considered by some to be a precursor of message and, since the length of nucleic acid tape required for polypeptide coding is relatively short, a large part of it is probably not structural message but may be a transcript of regulatory segments of the genome. If this model of message organization is at least approximately correct, chromosomes in higher organisms must have more DNA connected with regulating the expression of structural genes than with coding for specific proteins. Thus, even if foreign DNA coding for a particular protein is integrated into a genome in terms of replication and transmission, the question persists of how such DNA is subjected to effective regulation. Improper regulation could turn the dream of transformation into a nightmare, a possibility that Professor Ledoux has thus far avoided. And, since eukaryotes also have significant controls of biosynthesis at the translational level, instances of successful genetic transformation would represent a coordinated alteration in several tiers of intracellular regulation.

One other aspect of genetic transformation which should be considered is the mechanism by which heterologous DNA becomes integrated into a chromosome. If it is not so integrated it might exist as a

cytoplasmic particle or episome but, if so, its likelihood of being
transmitted to successive generations in concert with the nuclear
genome is rather small. In those cases where heterologous DNA is
known to be integrated into a chromosome, elucidation of the mechanism
would have broad biological implications. Chromosomes are highly
discriminating in selecting partners for recombination. This is obvious in meiotic systems where non-homologous chromosomes do not
undergo crossing over. Given the stringent conditions governing exchanges between DNA strands it would seem unlikely that transformation
by means of foreign DNA could be carried out with a substantial degree
of efficiency. Highly selective procedures are therefore needed to
recover a particular kind of transformation. The broader the spectrum
of sequence patterns present in the exogenous DNA, the broader must
be the spectrum of DNA sequences absorbed by a population of cells.
For each kind of foreign DNA molecule identified there must be many
others that remain undetected but are just as likely to effect heritable transformations. Thus, difficult though it may be for a particular foreign DNA molecule to become functionally integrated into a
genome, the relative number of other DNA molecules similarly integrated, - assuming the integration itself to be unselective, - is likely
to be large. The "directed" genetic transformation may thus have
little direction unless a purified gene is used.

These then are some of the problems that need to be overcome and
some of the hazards that may be encountered in seeking to bypass
natural channels for genetic exchange. A beginning appears to have
been made a number of years ago when Professor Ledoux first announced
his successful and highly efficient transformation of *Arabidopsis*.
Over the not so many years, the subject has been kept fresh by controversy. I'd like to call now upon Professor Ledoux to freshen the
subject a little more by providing the main address for this evening's
session.

DNA MEDIATED GENETIC CORRECTION OF
THIAMINELESS ARABIDOPSIS THALIANA

L. LEDOUX, R. HUART, M. MERGEAY, P. CHARLES

*Section de Biochimie cellulaire, Radiobiology Department,
Centre d'Etude de l'Energie Nucléaire, Mol, Belgium
and
Génétique Moléculaire, Universite dé Liège, Belgium*

and

M. JACOBS

*Laboratorium voor Plantengenetica, Vrije Universiteit Brussel,
Belgium.*

Results obtained by treating thiamineless mutants of
Arabidopsis thaliana with bacterial DNA indicate that the
genetic defect of the mutants can be corrected only with
DNA bearing a thiamine information. When the correction is
attempted in selective conditions, about 0.7% of the treated
plants grow and set fruit. This progeny and the following
ones, obtained by selfing, behave as homozygotes. The correction is hereditary. Results of back and test crosses
indicate that the correction is dominant, nuclear (and not
cytoplasmic) and strongly bound to the genome. The correction appears to be added to the mutation and not substituted for it: the correction behaves as an extragenic
suppressor.

It has been known that cell function in prokaryotes can be influenced by the addition of exogenous genetic information since Avery's
classical experiments. Subsequently it was demonstrated that bacteria
are capable of taking up DNA and integrating it into their genome by
genetic recombination.

Until recently, comparable "cell manipulations" with higher organisms appeared almost impossible, presumably due to the poverty of the
means available for sound and extensive studies. The necessary biochemical background had to be obtained; good methods to prepare nucleic acids and to recognize them in a population of similar molecules
had to be developed. At the genetical level, the scarcity of mutants
with defined biochemical lesions precluded (and still limits) experimentation (Ledoux, 1965; Bhargava and Shanmugan, 1971).

Nevertheless, DNA mediated corrections in higher organisms were re-

ported: with animal cells in tissue culture, (Szybalska and Szybalski,
1962; Glick and Salim, 1967; Majumdar and Bose, 1968; Fox *et al.*,
1969; Ottolenghi-Nightingale, 1969; Borenfreund *et al.*, 1970; Roosa, 1971; Merril *et al.*, 1971; Hill and Hillova, 1972); with
Ephestia (Nawa and Yamada, 1968; Nawa *et al.*, 1971); with Drosophila
(Gershenson, 1965; Fox and Yoon, 1966, 1970; Fox *et al.*, 1970, 1971a,
1971b); with Petunia (Hess, 1969, 1971, 1972, 1973) and with *Arabidopsis thaliana* (Ledoux and Jacobs, 1969; Ledoux and Huart, 1970;
Ledoux *et al.*, 1971a,b, 1972a,b, 1974).

Finally, phage mediate corrections were reported by Merril *et al.*
(1971) with animal cells, by Doy *et al.* (1972, 1973a,b,c) and Gresshoff and Doy (1972) with haploid tomato callus or Arabidopsis haploid
cells, and by Johnson *et al.* (1973) with Sycamore cells.

Extensive use of the CsCl gradient technique (first applied by
Gartler (1960) to the problem of DNA uptake by living cells) has led
to the astonishing results that foreign DNA can be translocated
large distances, cross cell membranes (Ledoux and Huart, 1972; Ledoux
et al., 1971c), and appear to survive several mitotic cycles in several animal and plant systems (Ledoux, 1965; Ledoux and Huart, 1972;
Ledoux and Charles, 1972a). In *Arabidopsis thaliana*, it appears to
survive at least one round of meiotic division (Ledoux *et al.*, 1971b,
1974; Brown *et al.*, 1971). More surprising was the suggestion that
the foreign DNA could become covalently linked to the recipient DNA
(Ledoux and Huart, 1970, 1972; Ledoux *et al.*, 1971b, 1972a; Laval
and Malaise, 1973) as a double stranded material (a phenomenon similar to the incorporation of the λDNA in the *E. coli* genome).

Since biochemical and biological data were generally obtained in
different laboratories with different materials, the *Arabidopsis
thaliana* system appeared to us as a useful one in which to try to
correlate biochemical and genetical data and to decide whether the
biochemical phenomena we observed were a kind of cul-de-sac or whether
they had implications for biology and genetics.

As already indicated in preliminary reports (Ledoux and Jacobs,
1969; Ledoux *et al.*, 1972a, 1974) corrections of nutritional mutants
by exogenous DNA were extensively developed in order to build the required statistical analysis; we are presenting here an overall picture of the results so obtained. More extensive genetic data will be
presented elsewhere.

NUTRITIONAL MUTANTS OF ARABIDOPSIS THALIANA.

Among the rare auxotrophs described in *Arabidopsis thaliana*, the thiamine requiring ones are the most frequent. They are obtained by X-ray or chemical mutagenesis. They also are conditional lethal mutants with a strict phenotype, whilst other auxotrophs exhibit bradytrophy.

One locus (py) controls the synthesis of the pyrimidine moiety of thiamine (Rédei, 1960, 1962, 1970), locus tz controls the synthesis of the thiazole half of the vitamin (Rédei, 1970) and 2 genes are con-cerned with some intermediate steps between these two precursors and thiamine. These mutants can be used for strictly genetic studies such

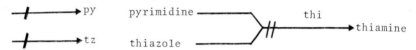

as allelic complementation or recombination or reverse mutation be-cause of the convenient selective mechanisms (Rédei, 1970). The mu-tability of the py locus is apparently the highest (Feenstra, 1965; Rédei, 1968) and is close to 2×10^{-4}. The frequency of spontaneous reversion has been estimated to be lower than 5×10^{-6} (Rédei, person-al communication). Ethylmethanesulphonate treatment produced appar-ent reversions at the py locus, at a frequency of about 4×10^{-5} (Van den Berg *et al.*, 1967). These revertants were heterozygous and segregated in the following generations. In test crosses, they also yielded heterozygotes and lethal plants in approximately equal fre-quencies (Feenstra and Van den Berg, 1968). No homozygous wild-types have been obtained by back mutation.

The monogenic conditional lethal mutants used here are normally propagated in the presence of thiamine. When sown on mineral medium, the seeds germinate normally and develop their cotyledons which appear normal green. The first leaves progressively bleach out completely and the plant soon dies. Thiamine (10^{-6}M - 10^{-5}M) allows normal growth of all these mutants.

Mutants used in this study are the following: thi51 and py er as, from Rédei, thi V447 and py V131 from Feenstra, and py 431 and tz 432 from Jacobs. They bear the side markers indicated in Tables 1 and 2. The only wild-type Arabidopsis kept in the laboratory is the Wilna gl.

TREATMENT OF THE SEEDS.

Seeds are disinfected with a 95% ethanol and 3% H_2O_2 mixture (1 : 1) as first used by Langridge (1955) or with a 5% Calcium hypo-

chlorite solution (Rédei, 1970). The seeds are treated for 5-8 mins
and rinsed with a few changes of sterile distilled water. About 50
seeds are then immediately placed in a well of a spot plate and im-
mersed in 25 µl of DNA solution (see later) or of 0.01 M NaCl. Spot
plates are placed in Petri dishes over wet sterile filter paper, to
limit evaporation of the solvent. During the following 3 days, 25 µl
of the DNA solution are added to the seeds to keep them moistened.
Petri dishes are kept at 24°C under continuous illumination for 4 days.
Both DNA and NaCl treated seeds germinate during the incubation period. After incubation, seedlings are laid out in a) tall Petri dishes,
on perlite moistened with the mineral nutrient solution of Jacobs
(1964) or b) in test tubes containing 5 ml 0.78% agar dissolved in
the mineral nutrient solution of Miksche and Brown (1965). The tubes
are placed in wooden racks and kept under continuous illumination
(6000 lux) at 24°C. All mutant seedlings on mineral medium die within
15-20 days; they show chlorophyll deficiency in cotyledons and rosette
leaves. Corrected plants grow at a slow rate, with a more or less
normal green colour. About one third of them produce fertile sili-
ques (see later). The thiamine concentration necessary to provide
complete growth when seeds were soaked for 3 days, then washed and
allowed to develop in the absence of thiamine, was found to be about
10^{-3}M.

DNA *PREPARATIONS*.

The series of DNA used is indicated in Table 1. They were prepared from frozen or fresh bacteria (or from isolated and CsCl purified bacteriophages) by the Marmur (1961) technique or by CsCl gradient preparative ultracentrifugation, after treatment with pronase and
RNase (Charles, 1972; Ledoux and Charles, 1972b). The DNA was precipitated with 2 vol. ethanol, kept in 70% ethanol, and dissolved in
sterile 0.01 M NaCl just before use. The concentration used was
about 0.5 - 1.0 mg/ml. We successively tried and rejected the use of
chloroform or phenol for deproteinization: both are toxic to the
plant. They also are known to remove the protein or membrane bound
molecules from the bulk of the DNA preparation.

Our study of the biochemical fate of labelled DNA in various systems led us to the recognition that: the DNA has to be freshly prepared, of high molecular weight (> 10^7 Daltons), very clean and intact. When single strand breaks in a DNA preparation could be detected from the depolymerising effects of alkali, the preparation was
discarded. Finally, the DNA had to be present during the first 24 hrs

Table 1

Corrections observed in DNA-treated populations

(no. corrected plants with a progeny/no. germinated seeds/no. treated seeds)

DNA	Lethal Mutants (additional markers)					Total
	thi 51 (gl)	thi V447 (er - gl)	tz 432 gl	py V131 (er, gl)	py 431 (-)	
A.						
E. coli	6/355/685	2/393/525	2/284/550	1/384/400	4/270/477	15/1686/2637
A. tumefaciens	2/182/420	0/100/100	3/230/480	1/100/200	0/130/250	6/742/1450
B. subtilis	0/128/288		3/157/328	0/28/100	0/162/420	3/497/1136
M. lysodeikticus	0/176/450		0/122/120		2/428/705	2/428/1375
S. coelicolor	0/145/274	1/75/75	0/152/272	0/22/100	0/120/410	1/524/1131
B.						
E. coli P678* (thi⁻A)			0/263/350	1/254/350		1/517/700
Phage T7 (E. coli)	0/425/542		0/440/500		0/356/520	0/1221/1562
Phage 2C (B. subt.)	0/189/252		0/182/220			0/371/472
C.						
NaCl 0.01 M	0/395/642	0/233/291	0/467/669	0/214/262	0/1580/2960	0/2889/5024

* requiring thiazole for growth.

of imbibition (otherwise it was not found integrated into the recipient material (unpublished data)).

In general, we used small amounts of seeds per batch of DNA, and
preferred to repeat the treatment with successive freshly prepared
batches. For some unknown reason, but supposedly due to the frag-
mentation of the DNA molecules during their isolation and purification, some batches of a given DNA species were ineffective with the
mutant plants tested, whilst others were quite effective. Yet, these
DNAs were indistinguishable on the basis of the physicochemical methods we used. Even if it is tempting to put the blame on the DNA
itself, it is of course difficult to decide whether these discrepan-
cies were not due to uncontrolled parameters related to plant growth.

The DNAs were prepared from the following strains: *E. coli* K12,
wild-type; *A. tumefaciens*, B6, obtained from Dr. A. Kurkdjian;
B. subtilis, 168 T$^+$; *M. lysodeikticus*, ATCC 4698; *S. coelicolor*
A3/2 obtained from Dr. D. Hopwood; *E. coli* P678 (thi A); *E. coli*
PA371 (thi A, recA) containing the KLF10 episome (thi A$^+$) kindly
supplied to us by Dr. Glansdorff. Phage T7 (with normal bases) and
Phage 2C (with hydroxymethyl uracil replacing thymine) were also
used. They were isolated and purified following the technique of
May *et al.* (1968).

CORRECTION IN THE DNA-*TREATED POPULATIONS.*

When seeds of the different mutant types were treated with different bacterial DNAs, corrections could be obtained with a relatively
high frequency comparable to that observed with bacterial transformation (10^{-2} to 10^{-4}). Tables 1 and 2 show an overall picture of the
data gathered during the last 3 years with a total of 27 different
batches of DNA given to five different mutants. These data include
DNA extracted from wild-type bacteria, from bacteria with a thiamine
deficiency and from phages containing normal or abnormal bases.
Further controls were obtained by treating mutant seeds with 0.01 M
NaCl, the solvent used for the DNA assays. After DNA treatment, a
few mutant seedlings, which normally die within 15 to 20 days after
sowing, display better growth, form small rosettes and eventually
flower. However, only a few of them are fertile (cf. Table 2).
Table 1 only takes into account the restored plants which produced
seeds, thus allowing a study of their progeny. Table 1 also indicates
in each case the number of germinating seeds and the total number of
seeds used. These results indicate that the percentage of correction
obtained with the group treated with DNA from wild-type bacteria is

Table 2

Results of treatments with *E. coli* DNA

				Lethal mutant		
Strain	py 431	tz 432	thi 51	py V131	thi V447	py
Additional mutation	-	- gl	- gl	er gl	er gl	er as gl
NaCl 0.01 M						
No. germinated seeds	1580	467	395	214	233	194
No. developed plants	0	0	1	0	0	0
No. fertile plants	0	0	0	0	0	0
% correction	0	0	0	0	0	0
***E. coli* thi⁺**						
No. germinated seeds	270	284	335	384	293	282
No. developed plants	10	5	12	3	5	3
No. fertile plants	4	2	6	1	2	0
% correction	1.14	.70	1.69	.26	.51	0
***E. coli* thi⁻**						
No. germinated seeds		263		254		194
No. developed plants		2		1		0
No. fertile plants		0		1		0
% correction		0		.39		0
***E. coli* thi⁻ + episome thi⁺**						
No. germinated seeds	72	96	93		94	93
No. developed plants	16	12	17		2	18
No. fertile plants	4	1	1		0	0
% correction	6.15	1.04	1.11		0	0

similar for all *Arabidopsis* mutants: 0.7%, 0.85% and 0.49% respectively for thiamine, thiazole and pyrimidine requiring mutants. On the other hand, the percentages of correction obtained for five mutants seem to vary with the source of DNA, *E. coli* being at the moment the more effective one (0.89%) and *S. coelicolor* the least effective DNA (0.19%). Results obtained with DNA from *E. coli* thi⁻, phage T7 or phage 2C were all negative.

However, when a thi⁻ bacteria harbours a F' thi⁺ episome, correction is obtained. *E. coli* PA371 used as a recipient was derived from the P678 strain; strain PA 371 carries recA, which prevents the episome recombining with its genome. The DNA extracted from such a PA 371 + KLF10 bacteria contained about 95% PA371 DNA and 5% episomal DNA, in free form. Table 3 summarizes the results obtained with six Arabidopsis mutants and the different *E. coli* DNA used. Correction is found with the DNA extracted from the thi⁻ bacteria harbouring the episome. As the corresponding bacterial chromosomal DNA was inefficient, the corrections obtained should be due to the episomal information.

Table 3

Summary of results of the DNA-treatment

Treatment

	NaCl 0.01 M	Phage or *E. coli* DNA without thiamine information 0.6 - 1.0 mg/ml 0.01 M NaCl	DNA from thiamine⁺ bacteria 0.6-1.0 mg/ml 0.01 M NaCl
No. treated seeds	5,024	2,634	7,729
No. germinating seeds	2,889	2,033	3,877
No. developed plants	1	2	73
No. fertile plants	0	0	27

It should also be noted that in all cases, the DNA treated plants growing in the absence of thiamine show a delay in growth and development, as compared to the wild-type or to mutants supplemented with thiamine (5×10^{-5}M). In two cases, white segments were apparent on the stems of the corrected plants, indicating mosaicism and thiamine deficiency.

ANALYSIS OF THE PROGENIES OBTAINED BY SELFING THE CORRECTED TYPES.

Progenies of various corrected types obtained after DNA treatment were sown in perlite soaked with mineral medium or in soil, and checked during the entire growth of the plant. The number of F_1 tested plants varied as a function of the number of seeds present in the pods of the corrected plant growing on a thiamine-less medium. In a very few cases, thiamine was provided at the flowering stage to increase the vigour and thus the seed production. F_2 and F_3 progenies include at least fifty plants for each experiment. For most corrected plants, up to three successive generations were studied and for 5 corrected mutants, fresh and dry weights were determined (Table 4).

The most striking point of this study concerns the lack of segregation upon selfing the progenies of the various corrected types. All F_1 to F_3 plants originating from each corrected plant look phenotypically alike and grow reasonably well on mineral medium. This was also observed in one case where the selfing was pursued up to the 7th generation. They show however, in most - but not all cases - a variegated pattern of chlorophyll pigmentation with light discoloration of the basis of the limb and of the region along the main vein. Practically in all tested cases, the average fresh weight (and particularly the average dry weight) of the corrected types grown on mineral medium, are lower than those of plants from the same progeny, supplemented with 5×10^{-4}M thiamine (44 to 92%) (Table 4). Each of these types thus displays sensitivity to the addition of thiamine. Other experiments showed that the corrected "pyrimidine" or "thiazole" mutants specifically respond to the addition of pyrimidine or thiazole and show an improved growth.

The offspring of mutants corrected in 1970 were again tested in 1973. Despite a larger number of non germinating seeds, results were essentially the same and no segregation could be observed among the growing plants. In selfing experiments, the correction thus appears to be stable with time. The corrected plants behave as true breeding homozygotes in contrast to what is found with mutagen induced revertants (Ledoux and Huart, 1970). This is also in contrast to what is expected in the case of a suppressor mutation introduced in a homozygous diploid. Indeed, in the seed the germline is represented by two diploid cells (Langridge, 1958; Li and Rédei, 1969; Nikolov and Ivanov, 1969; Ivanov, 1971). Both these cells should contain the thiamine information to explain the lack of segregation

Table 4

Ratios of the weights of F_2 progenies

(obtained from 5 corrected mutants or from the wild-type)

grown on MM, to the weight observed after growth on MM + thiamine 5.10^{-4}M

(21 days culture) (results MM + thiamine = 100)

Genotype of the treated plants	Correcting bacterial DNA	Phenotype of the corrected plant	No. of tested plants	Fresh weight	Dry weight
py 431	M. lysodeikticus	light green, va-riegations	27	102.9	85.4
	M. lysodeikticus	normal green, va-riegations	13	91.8	73.9
tz 432	E. coli	normal green, no variegation	24	78.2	81.5
	A. tumefaciens	light green, va-riegations	29	43.8	43.8
	B. subtilis	light green, no variegation	13	93.5	91.8
wild-type		normal green, no variegation	22	107.8	105.8

after selfing. In F_2, the ratio of mutants : wild-types should vary
from 1 : 3 to 1 : 7. The fact that correction ends up with an homozygous corrected plant suggests that there has been a continuous selection for the corrected germ cells and, presumably a meiotic drive
at the time of meiosis favouring those gametes which harbour the thiamine information. Indeed, when the selection pressure is prevented
by thiamine supplementation of DNA-treated plants showing signs of
abortive flowering, a high percentage of lethal mutants is observed
in the F_1 progeny.

ANALYSIS OF THE OFFSPRING OF CROSSES.

We have analysed, in parallel, the offspring of crosses made between corrected plants and either the wild or the original mutant
type. Tables 5 and 6 show the results obtained in the case of tz⁻,
an homozygous "thiazole" mutant corrected by a treatment with
B. subtilis DNA. (Similar data were and are being gathered for other
corrected types such as py^c, th^c or tz^c corrected with other DNA.
They will be described elsewhere together with the results of crosses
between mutants corrected with different DNA). The phenotypes presented by progenies of successive self-fertilization were analysed.

Table 5
Results of a cross between tz 432,
lethal, corrected by *B. subtilis* DNA, and the wild-type

Cross	Progeny tested	Year	No. growing plants	No. ungermin. seeds	% Phenotypes		
					Normal	Leaky	Lethal
tz^c x tz^+	X →XF_1	1971	11	20	100		
	XF_1 →XF_2	1971	163	15	90.1	9.2	0.7
	(XF_{2N} →XF_3	1971	198	5	95.5	2.0	2.5
	(XF_{2Lk}→XF_3	1971	198	13	21.7	70.2	8.1
	XF_1 →XF_2	1973	292	145	80.1	13.0	6.9
	(XF_{2N} →XF_3	1973	138	15	84.8	11.6	3.6
	(XF_{2Lk}→XF_3	1973	57	11	5.3	17.5	77.2

The first generation of the outcrosses do not segregate (Table 5
and 6). The results of the test cross (Table 6) imply that the correction is dominant and that the corrected plants behave as true homozygotes. Reciprocal crosses between corrected mutants and wild-type

Table 6

Results of a cross between a tz 432,

lethal, corrected by *B. subtilis* DNA, and the tz 432 mutant

Cross	Progeny tested	Year	No. growing plants	No. ungermin. seeds	% Phenotypes		
					Normal	Leaky	Lethal
tz^c x tz^-	X $\rightarrow XF_1$	1971	20	5	100		
	$XF_1 \rightarrow XF_2$	1971	116	15	75.8	6.1	17.3
	($XF_{2N} \rightarrow XF_3$	1971	131	4	76.3	0	23.7
	($XF_{2Lk} \rightarrow XF_3$	1971	202	21	24.7	65.0	10.3
	$XF_1 \rightarrow XF_2$	1973	176	251	59.7	12.5	27.8
	($XF_{2N} \rightarrow XF_3$	1973	99	3	72.7	19.2	8.1
	($XF_{2Lk} \rightarrow XF_3$	1973	109	38	2.8	27.5	69.7

plants or lethal mutants (cf. Table 7) indicated that the correction
can be transmitted through the male as well and through the female
gamete and therefore does not appear to be due to a maternal or cytoplasmic factor.

Table 7

F_1 progeny of reciprocal crosses between

mutants and corrected mutants

Cross	No. growing plants	No. ungerm. plants	% phenotypes		
			Normal	Leaky	Lethal
tz^c x tz	36	4	36	-	-
tz x tz^c	17	5	17	-	-
py^c x py	51	10	51	-	-
py x py^c	15	15	15	-	-
thi^cx thi	184	50	184	-	-
thi x thi^c	49	39	56	-	-

As exemplified in Table 5, the crosses tz^c x tz^+ always lead to
high percentages of non-germinating seeds (60%-75%) much higher than
the percentages observed in crosses tz^c x tz^- (15%-20%) or tz^- x tz^+
(0%-5%) (Table 8). This suggests a sort of deleterious interaction
between the correcting and the original thi^+ information.

Table 8-
Results of a cross between
the tz 432 mutant and the wild-type

Cross	Progeny tested	Year	No. growing plants	No. ungermin. seeds	% Phenotypes		
					Normal	Leaky	Lethal
$tz^- \times tz^+$	$X \rightarrow XF_1$	1971	15	0	100		
	$XF_1 \rightarrow XF_2$	1971	106	7	72.6	0	27.4
	$XF_1 \rightarrow XF_2$	1973	63	4	73.0	0	27.0

Segregation is observed in the later generation offspring of the $tz^c \times tz^+$ cross. Lethal mutant types are recovered (although in low frequencies) together with plants exhibiting a new phenotype called here "leaky mutant". This is characterized by a weaker pigmentation, white patches of discoloration on the leaves, sometimes white sectors on the stem and a relatively poor growth and fertility, in contrast with the normal looking plants of the same progeny. A whole range of these phenotypes intermediate between the normal green plants and the lethal chlorophyll deficient plants were classified under the name "leaky mutants". These leaky mutants are obtained in crosses between the corrected type and the wild-type, as well as in the crosses between the corrected type and the lethal mutant (in the latter case, a high frequency of lethal mutants being obtained) (Tables 5 and 6). The leaky types retain their sensitivity to supplementation with thiamine, thiazole or pyrimidine respectively. This indicates that the correction has been added to the mutation and has not been substituted for it. Correction being dominant, the recessive mutation is not expressed unless the correction is masked or lost, for instance by crossing with an uncorrected plant.

When such progenies are tested after two years storage, the percentages found differ considerably ($XF_1 \rightarrow XF_2$, 1971 vs. 1973). The percentage of non-germinating seeds is high in both $tz^c \times tz^+$ and $tz^c \times tz^-$ crosses. This might be due to the instability of a genetic factor linked to the correction, rather than to a physiological factor. Indeed, storage of XF_1 seeds from a cross $tz^- \times tz^+$ does not affect the germination of XF_2 (Table 8). It appears that embryonic lethality could be the reason for this phenomenon. In fact, these non germinating plants cannot be stimulated by thiamine nor by yeast extract. Microscopical examination reveals no abnormality and in some cases a growing but abortive meristematic zone can be observed in these embryos.

Besides this increased percentage of non germinating seeds, the percentage of leaky and lethal mutants obtained drastically increases in both crosses. The phenotypically normal looking plants of the XF_1 progenies do segregate in their progenies (XF_2N) for lethal and leaky types (XF_2Lk). Storing the seeds for two years increases the tendency to segregate. It can therefore be concluded that the correction, stable with time in selfing experiments (see above) becomes unstable with time upon crossing DNA-corrected plants with normal wild or mutant types. It is thus clear that the cross tz^c x tz^+, bringing together two phenotypically thi^+ partners, leads to an important segregation pattern, never observed when tz^c is self fertilized.

Let us now consider more closely the "leaky mutants" obtained. They are found in equal frequencies in the XF_1 offsprings from tz^c x tz^+ (9.2%) and tz^c x tz^- (6.1%). Their offpsring include high frequencies of leaky and lethal types as well as normal looking plants. The normal phenotype can be recovered at high frequencies (21.7% and 24.7%) suggesting that the "leaky mutants" behave as heterozygotes. They should however also segregate lethal mutant types with similar frequencies. The lower values obtained for the lethal mutants and the reproducibility of the phenotype distributions could be accounted for by various mechanisms.

We expected that the present study could help in correlating genetical and biochemical data. This was in fact the main reason why bacterial DNA was used in the biological studies. We have not used homologous *Arabidopsis* DNA owing to the difficulty of preparing from this material the clean, high molecular weight and intact DNA considered by us as a prerequisite for efficient uptake and integration (cf. above). On the other hand, whilst we wanted to use specialised transducing phages to introduce selected genes only[*], we were not in a position to do so, because of the lack of phages carrying thiamine

---

[*]The amount of DNA which can be taken up by a cell is limited as are the sizes of the populations handled. It therefore seems reasonable to try to increase the dosage of the interesting gene in the DNA preparation used. This can be achieved by using DNA from bacteria, transducing phages or episomes.

information. This kind of difficulty has since been overcome by using DNA from thiamineless bacteria (ineffective as such for correction) harbouring a F' factor bearing the thiamine information (KLF10).

Analyses in CsCl gradients have shown that seeds do accumulate large amounts of heteropycnic foreign DNA. In fact, dry seeds swelling in a DNA preparation (1 mg/ml) absorb about 0.7 ng DNA per seed (Ledoux et al., 1971c). The embryo contains about 10^4 cells, with 0.8 pg DNA per cell, so that the amount of foreign DNA taken up by each cell is equivalent to about 20 E. coli genomes and corresponds to 1/10 of the actual DNA content of the cell. This huge amount of DNA thus taken up by a cell contains on average, 20 thiamine genes. One plant (i.e. one germ cell) in 100 being corrected, the efficiency of the correction, at the molecular level, is about 10^{-4}. The 1% efficiency observed for the whole organism is due to the biological amplification afforded by the system used.

In the embryo, the foreign DNA is mainly stored in the cotyledons and remains associated with the cotyledon DNA for most of the plant growth. At flowering time, when cotyledons senesce, the foreign DNA seems to be excised and migrates to the flowers from where it is transferred to the seeds[**]. These results suggest that the foreign DNA can act as an "episome resembling" DNA, being found either integrated or free.

Autoradiography on the other hand, shows that the foreign DNA becomes associated with cell nuclei, migrates at the time of flowering through the stem vessels and is found associated with pollen grains. In the flowers of plants treated with M. lysodeikticus DNA and ^{3}H thymidine the foreign integrated DNA replicates; it also replicates in the progeny (Ledoux et al.,1971b; Ledoux et al., 1972a). At this point the biochemical picture could be correlated with the genetic behaviour of the corrected mutants. A Micrococcus-Arabidopsis satellite peak is observed, the importance of which increases upon treatment of successive generations of plants with the same bacterial DNA (Brown et al., 1971; Ledoux et al., 1971b; Ledoux et al., 1972a) This is accompanied by a series of biological modifications, such as a drastic decrease of the % germination, a delay or a complete suppression of the flowering or the appearance of mutant like phenotypes

[**]This pattern of DNA release is also found in Sinapis alba (Coumanne et al., 1971).

(hairy plants, rosula shape of the rosette, browning of the leaves, white or dark brown seeds, etc.). All these effects point toward a DNA-induced imbalance of the plant metabolism.

To what type of genetical model do these results lead? Different models have been discussed by Fox *et al.* (1971a) to explain their results showing genetic transformation in *Drosophila* with homologous DNA. They argue in favour of the exosome model, also considered by Hess (1971) in interpreting the transformation of Petunia. The exosome model (Fig 1) is the only one which does not require an integration of the exogenous information into the linear structure of the chromosome. Such an integration is rejected by Fox due to the absence of whole-body transformants in *Drosophila* and to the instability of the correction. In the *Arabidopsis* system, the non-autonomous type of biochemical lesions due to the possible diffusion of thiamine, makes it difficult to ascertain the predicted mosaic nature of the corrected type. Its progeny however is easier to analyse. The corrected plants behave as homozygotes as their offspring do not segregate (they also do not segregate 1 to 1 in test crosses). As correction is stable in subsequent offspring, it would correspond to a surprisingly stable form of exosome. The effect of the exogenous DNA appears to be associated with the concerned gene (thiamine, thiazole or pyrimidine loci). This leads to the concept of an association of the bacterial information with the *Arabidopsis* genome. Such a close association is also indicated by the absence of significant differences between progenies of reciprocal crosses. The thiamine mutation remains present in the genome of the corrected type; by crossing the corrected plants with the wild-type, the original mutant type does reappear in F_2 although at very low frequencies. Passage in plants arising from a cross with one of the corrected types through meiosis seems to increase the instability of the exogenous information. One of the consequences is the appearance of the "leaky mutant" phenotype (which could be considered as functional mosaics). The correction also becomes unstable with time; storing F_1 or F_2 seeds resulting from crosses between corrected and mutant or wild-types, results in the appearance of an increased proportion of mutants. The regularity in the transmission of the genetic information as well as the survival of the mutated site in the corrected plant, could be interpreted if we assume that the information has been added to the genome (and not substituted for the mutation) and can be removed by crossing the plant with the wild-type or with the mutant but not upon selfing. Difficulties in

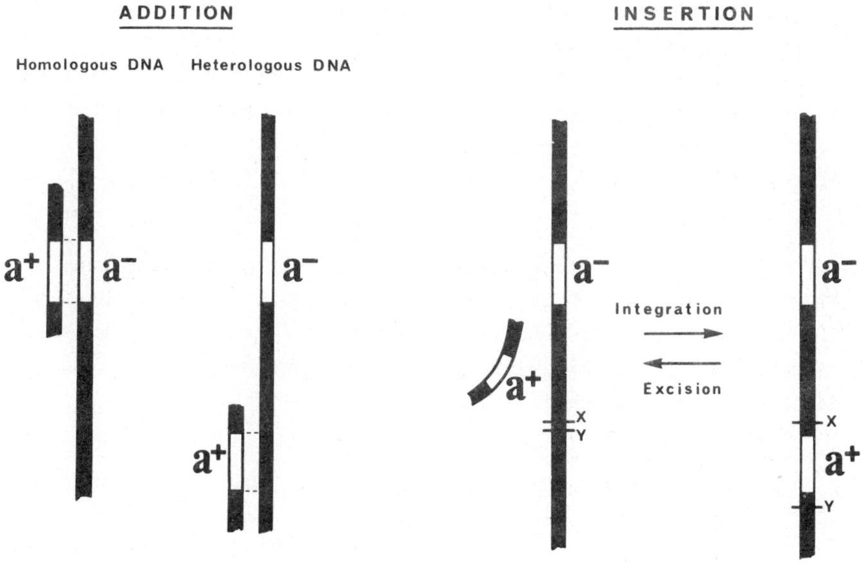

Fig 1 Comparison of addition and insertion models (one of two homologous chromosomes is shown).

In *Drosophila*, transformed by homologous DNA, Fox's exosome model
implies a close association of the foreign gene with its homologous
chromosome locus. This exosome is not integrated but replicates in
step with the chromosome and can be lost or transmitted with it at
the time of cell division. Either the exosomal gene or the chromosomal one is transcribed, leading to phenotypic mosaicism.

When heterologous DNA is used for correction, as in our case, the
necessity of a close association with the mutated site is less manda-
tory and the heterologous exosome could become associated to another
region, where the actual base sequence is in favour of such an association. Here too, the "exosome" could replicate in step with the
chromosome and be lost or transmitted with it at the time of cell
division. The probability of loss would however be high due to the
imperfect homology.

In the insertion model, the foreign gene is integrated in the recipient DNA, like episomal DNA, through the homology of a few base
pairs. Once integrated, it would be hard to lose, except upon pairing with a chromosome homologous for the rest of the structure but
lacking the inserted foreign piece. Whilst gene conversion could
interfere at that step, chromosome aberration is an obvious alternative, possibly leading to the loss of the foreign gene.

chromosome pairing, due to the correction present in only one of the
two homologous chromosomes forming the bivalent, could be responsible
for the removal of this correction. Such a picture fits well with

the "episomal DNA" type of interpretation (cf. Fig 1) emerging from the biochemical analysis. This model also provides an explanation for the variability in the functioning of the bacterial thiamine gene in the recipient plant, possibly related to its relative position. These differences in the phenotypic expression of the thiamine locus do not appear to be due to a lack of integration but to a change in the gene environment able to influence its genetic transcription and therefore its expression.

ACKNOWLEDGEMENTS

We thank J. Swinnen-Vranckx and L. De Mol for their excellent technical help and Drs. Glansdorff and P. Lurquin for numerous fruitful discussions. We are indebted to the Fonds National de la Recherche Fondamentale Collective and to the Ministère de l'Education Nationale for their financial help.

REFERENCES

Avery, O.T., McLeod, C.M., and McCarthy, M. (1944). Studies on the chemical nature of the substance inducing transformation of pneumococcal types. Induction of transformation by a DNA isolated from pneumococcus type III. *Journal of Experimental Medicine.* 79, 137-157.

Bhargava, P.M., and Shanmugan, G. (1971). Uptake of non viral nucleic acids by mammalian cells. In Progress in Nucleic Acid Research and Molecular Biology (Edit. by J.N. Davidson and W.E. Cohn). Academic Press, New York. 11, 103-292.

Borenfreund, E., Honda, Y., Steinglass, M., and Bendich, A. (1970). Studies of DNA-induced heritable alteration of mammalian cells. *Journal of Experimental Medicine.* 132, 1071-1089.

Brown, J., Huart, R., Ledoux, L., and Swinnen-Vranckx, J. (1971). High specific radioactivity of a heavy DNA fraction in progeny of Arabidopsis treated by *Micrococcus lysodeikticus* DNA. *Archives Internationales de Physiologie et de Biochimie.* 79, 820-821.

Charles, P. (1972). Isolation, preparation and characterization of Deoxyribonucleic Acids. In Uptake of Informative Molecules by Living Cells (Edit. by L. Ledoux). (North-Holland). 10-28.

Coumanne, C., Jacqmark, A., Kinet, J.M., Bodson, M., Ledoux, L., and Huart, R. (1971). Translocation and distribution of bacterial DNA in *Sinapis alba*. *Archives Internationales de Physiologie et de Biochimie.* 79, 823-824.

Doy, C.H., Gresshoff, P.M., and Rolfe, B.G. (1972). Transfer and expression (transgenosis) of bacterial genes in plant cells.

Search. 3, 447-448.

Doy, C.H., Gresshoff, P.M., and Rolfe, B.G. (1973a). Biological and molecular evidence for the transgenosis of genes from bacteria to plant cells. *Proceedings of the National Academy of Sciences, U.S.A.* 70, 723-726.

Doy, C.H., Gresshoff, P.M., and Rolfe, B.G. (1973b). Transgenosis of bacterial genes from *E. coli* to cultures of haploid *Lycopersicon esculentum* and haploid *Arabidopsis thaliana* plant cells. In The Biochemistry of Gene Expression in Higher Organisms (Edit. by J. Pollak and J. Wilson Lee). (Australia and New Zealand Book Co.). 21-37.

Doy, C.H., Gresshoff, P.M., and Rolfe, B.G. (1973c). Time course of phenotypic expression of *E. coli* gene Z following transgenosis in haploid *Lycopersicon esculentum* cells. *Nature New Biology.* 244, 90-91.

Feenstra, W.J. (1965). Production of thiamineless mutants. *Arabidopsis Information Service.* 2, 25.

Feenstra, W.J., and Van den Berg, B.I. (1968). Continuously segregating revertants of pyrimidineless mutants of *Arabidopsis thaliana*. *Proceedings XII International Congress of Genetics.* 1, 26.

Fox, M., and Ayad, S.R. (1972). Uptake and integration of exogenous DNA by lymphoma cells. In Uptake of Informative Molecules by Living Cells (Edit. by L. Ledoux). (North-Holland). 295-312.

Fox, A.S., Duggleby, W.F., Gelbart, W.M. and Yoon, S.B. (1970). DNA-induced transformation in *Drosophila:* evidence for transmission without integration. *Proceedings of the National Academy of Sciences, U.S.A.* 67, 1834-1838.

Fox, M., Fox, B.W., and Ayad, S.R. (1969). Evidence for genetic expression of integrated DNA in lymphoma cells. *Nature.* 222, 1086-1087.

Fox, A.S., and Yoon, S.B. (1966). Specific genetic effects of DNA in *Drosophila melanogaster. Genetics.* 53, 897-911.

Fox, A.S., and Yoon, S.B. (1970). DNA-induced transformation in *Drosophila:* locus-specificity and the establishment of transformed stocks. *Proceedings of the National Academy of Sciences, U.S.A.* 67, 1608-1615.

Fox, A.S., Yoon, S.B., Duggleby, W.F., and Gelbart, W.M. (1971a). Genetic transformation in *Drosophila:* In Informative Molecules in Biological Systems (Edit. by L. Ledoux). (North-Holland). 313-333.

Fox, A.S., Yoon, S.B., and Gelbart, W.M. (1971b). DNA-induced transformation in *Drosophila:* genetic analysis of transformed stock. *Proceedings of the National Academy of Sciences, U.S.A.* 68, 342-346.

Gartler, S.M. (1960). Demonstration of cellular uptake of polymerized DNA in mammalian cell cultures. *Biochemical and Biophysical Research Communications*. 3, 127-131.

Gershenson, S.M. (1965). Mutagenic action of some biopolymers in *Drosophila*. *Genetical Research*. 6, 157-162.

Glick, J.L., and Salim, A.P. (1967). DNA-induced pigment production in a hamster cell line. *Journal of Cell Biology*. 33, 209-212.

Gresshoff, P.M., and Doy, C.M. (1972). Development and differentiation of haploid tomato. *Planta*. 107, 161-170.

Hess, D. (1969). Versuche zur transformation an höheren pflanzen: Wilderholung der anthocyan-induktion bei Petunia und erste charakterisierung des transformierenden prinzips. *Zeitschrift für Pflanzenphysiologie*. 60, 348-358.

Hess, D. (1970). Versuche zur transformation an höheren pflanzen: mogliche transplantation eines gens für blattform bei *Petunia hybrida*. *Zeitshcrift für Pflanzenphysiologie*. 63, 461-467.

Hess, D. (1972). Transformationen an höheren organismen. *Naturwissenschaften*. 59, 348-355.

Hess, D. (1973). Transformationsversuche an höheren pflanzen: untersuchungen zur realisation des exosomen-modells der transformation bei *Petunia hybrida*. *Zeitschrift für Pflanzenphysiologie*. 68, 432-440.

Hill, M., and Hillova, J. (1972). Recombinational events between exogenous mouse DNA and newly synthesized DNA strands of chicken cells in culture. *Nature*. 237, 35-39.

Hotta, Y., and Stern, H. (1971). Uptake and distribution of heterologous DNA in living cells. In Informative Molecules in Biological Systems (Edit. by L. Ledoux) (North-Holland). 176-186.

Ivanov, V.I. (1971). Leaf colour mutants in *Arabidopsis* induced by gamma-irradiation of dormant seeds. *Arabidopsis Information Service*. 8, 29.

Jacobs, M. (1964). Nutrient recipes for selection and growing of biochemical mutants. *Arabidopsis Information Service*. 1, 36.

Johnson, C.B., Grierson, D., and Smith, H. (1973). Expression of λ plac 5 DNA in cultured cells of a higher plant. *Nature New Biology*. 244, 105-107.

Langridge, J. (1955). Biochemical mutations in the crucifer *Arabidopsis thaliana*. *Nature*. 176, 260-261.

Langridge, J. (1958). A hypothesis of developmental selection exemplified by lethal and semi-lethal mutants of *Arabidopsis*. *Australian Journal of Biological Sciences*. 11, 56-68.

Laval, F., and Malaise, E. (1973). Analysis of steps in penetration
    and integration of bacterial DNA in normal or X-irradiated mice
    fibroblasts. In Bacterial Transformation (Edit. by L.J. Arches).
    (Academic Press). 387-405.
Ledoux, L. (1965). Uptake of DNA by living cells. In Progress in
    Nucleic Acid Research and Molecular Biology (Edit. by J.N. David-
    son and W.E. Cohn). (Academic Press, New York). 231-267.
Ledoux, L., Brown, J., Charles, P., Huart, R., Jacobs, M., Remy, J.,
    and Watters, C. (1972a). Fate of exogenous DNA in mammals and
    plants. In Advances in the Biosciences (Edit. by G. Raspé).
    (Pergamon Press). 8, 347-367.
Ledoux, L., and Charles, P. (1972a). Fate of exogenous DNA in mammals.
    In Uptake of Informative Molecules by Living Cells (Edit. by L.
    Ledoux). (North-Holland). 397-413.
Ledoux, L., and Charles, P. (1972b). On the use of preparative CsCl
    gradients. In Uptake of Informative Molecules by Living Cells
    (Edit. by L. Ledoux). (North-Holland). 29-46.
Ledoux, L., and Huart, R. (1970). Fate and possible role of exogenous
    bacterial DNA in barley. In Barley Genetics (Edit. by R.A. Nilan).
    (Washington University Press). 2, 254-263.
Ledoux, L., and Huart, R. (1972). Fate of exogenous DNA in plants.
    In Uptake of Informative Molecules by Living Cells (Edit. by L.
    Ledoux). (North-Holland). 254-276.
Ledoux, L., Huart, R., and Jacobs, M. (1971a). Etude des descendants
    de plantes d'Arabidopsis thaliana traitees par les DNA bacteriens.
    Archives Internationales de Physiologie et de Biochimie. 78, 591-
    593.
Ledoux, L., Huart, R., and Jacobs, M. (1971b). Fate of exogenous DNA
    in Arabidopsis thaliana. In Informative Molecules in Biological
    Systems (Edit. by L. Ledoux). (North-Holland). 159-175.
Ledoux, L., Huart, R., and Jacobs, M. (1971c). Fate of exogenous DNA
    in Arabidopsis thaliana. European Journal of Biochemistry. 23,
    96-108.
Ledoux, L., Huart, R., and Jacobs, M. (1972b). Fate and biological
    effects of exogenous DNA in Arabidopsis thaliana. In The Way Ahead
    in Plant Breeding (Edit. by F.G. Lupton, G. Jenkins and R. Johnson).
    (Adlard & Son, Dorking). 165-184.
Ledoux, L., Huart, R., and Jacobs, M. (1974). DNA-mediated genetic
    correction of thiaminelessArabidopsis thaliana. Nature. 249, 17-21.
Ledoux, L., and Jacobs, M. (1969). Redistribution lors de la florai-
    son des DNA exogenes absorbes par des graines d'Arabidopsis thali-

88

ana. Archives Internationales de Physiologie et de Biochimie.
77, 568-569.

Li, S., and Rédei, G.P. (1969). Estimation of mutation rate in auto-
gamous diploids. *Radiation Botany.* 9, 125-131.

Majumder, A., and Bose, S.K. (1968). DNA mediated genetic transformation of a human cancerous cell line cultured *in vitro. British
Journal of Cancer.* 22, 603-613.

Marmur, J. (1961). A procedure for the isolation of Deoxyribonucleic
Acid from micro-organisms. *Journal of Molecular Biology.* 3, 208-
218.

May, P., May, E., Granboulan, P., Granboulan, N., and Marmur, J.
1968). Ultrastructure de bactériophage 2. C et propriétés de son
DNA. *Annals de l'Institut Pasteur.* 115, 1029-1031.

Merril, C.R., Geier, M.R., and Petricciani, J.C. (1971). Bacterial
virus gene expression in human cells. *Nature.* 233, 398-400.

Miksche, J.P., and Brown, J.A.M. (1965). Development of vegetative
and floral meristems of *Arabidopsis thaliana. American Journal of
Botany.* 52, 533-537.

Nawa, S., Sakaguchi, B., Yamada, M.A., and Tsujita, M. (1971). Hereditary change in Bombyx after treatment with DNA. *Genetics.* 67,
221-252.

Nawa, S., and Yamada, M.A. (1968). Hereditary change in Ephestia after
treatment with DNA. *Genetics.* 58, 573-584.

Nikolov, C.V., and Ivanov, V.I. (1969). Effect of heat treatment
and gamma irradiation of the seeds of *Arabidopsis thaliana* (L.)
Heynh. on the mutation rate in M_2. *Genetika.* 5, 168-170.

Ottolenghi-Nightingale, E. (1969). Induction of melanin synthesis in
albino mouse skin by DNA from pigmented mice. *Proceedings of the
National Academy of Sciences, U.S.A.* 64, 184-189.

Rédei, G.P. (1960). Genetic control of 2,5-dimthyl-4-aminopyrimidine
requirement in *Arabidopsis thaliana. Genetics.* 45, 1007.

Rédei, G.P. (1962). Genetic block of vitamin thiazole synthesis in
Arabidopsis. Genetics. 47, 979.

Rédei, G.P. (1968). A comparison of the somatic effect of X-rays and
ethyl methanesulfonate. *Arabidopsis Information Service.* 5, 36.

Rédei, G.P. (1970). *Arabidopsis thaliana.* A review of the genetics
and biology. *Bibliographia Genetica.* 20, 1-151.

Relichova, J. (1972). The distribution of induced mutations in the
shoot apex of *Arabidopsis* seeds. *Arabidopsis Information Service.*
9, 28-29.

Roosa, R.A. (1971). Induced and spontaneous metabolic alterations in
   mammalian cells in culture. In Informative Molecules in Biological
   Systems (Edit. by L. Ledoux). (North-Holland). 67-79.

Szybalska, E.H., and Szybalski, W. (1962). Genetics of human cell
   lines IV. DNA-mediated heritable transformation of a biochemical
   trait. *Proceedings of the National Academy of Sciences, U.S.A.*
   48, 2026-2034.

Van den Berg, B.I., Heyting, J., and Feenstra, W.J. (1967). Revertants
   of pyrimidineless mutants. *Arabidopsis Information Service.* 4, 46.

STUDIES ON THE UPTAKE AND EXPRESSION OF
FOREIGN GENETIC MATERIAL BY HIGHER PLANT CELLS

D. GRIERSON, R.A. McKEE, T.H. ATTRIDGE AND H. SMITH

*Department of Physiology and Environmental Studies,
University of Nottingham School of Agriculture,
Sutton Bonington, Loughborough. LE12 5RD.*

When sycamore cells grown in suspension culture on mineral
salts and sucrose are treated with λ p *lac* 5 transducing
phage, plaque-forming particles disappear from the supernatant and the cells are subsequently able to grow and divide
slowly on a medium containing lactose in place of sucrose.
Untreated cells and those treated with λ^+ phage are unable
to grow on lactose. λ p *lac* 5-treated cells grow normally
on sucrose but only very slowly on lactose. Typically there
is a burst of cell division after a few weeks but this is
not maintained and the number of cells in the culture eventually decreases. The failure of the cells to continue growing on lactose cannot be explained by the presence of galactose in the culture medium.

There is considerable evidence that whole plants and cultured plant
cells are capable of taking up purified DNA although there is some controversy concerning the fate of exogenous DNA. Part of it may be broken down and reutilised for host DNA synthesis but some workers have
suggested that it can also become covalently linked to host DNA without prior degradation (see Johnson and Grierson, 1974, for a short review). The apparent corrections of metabolic deficiencies by exogenous DNA could be explained by the uptake and expression of the appropriate genetic information (Ledoux *et al.*, 1974). There is no doubt
that such explanations are valid for experiments with bacteria, and
detailed studies on nucleic acid metabolism and protein synthesis will
establish whether this is also the case for plants. For this purpose
it is obviously desirable to carry out experiments with donor DNA containing defined nucleotide sequences so that these and the enzymes
they specify may be positively identified in recipient cells. We have
used a *lac* transducing phage, λ p *lac* 5, constructed by Ippen *et al.*
(1971) as a donor for the gene coding for the *Escherichia coli* β-gala-
tosidase and have investigated the properties of cell suspension cultures of sycamore (*Acer pseudoplatanus*) treated with the phage.

RESULTS

GROWTH OF SYCAMORE CELL SUSPENSION CULTURES.

The pattern of cell division in suspension cultures of *Acer pseudo-platanus* is shown in Figure 1. With sucrose as carbon source cell number rapidly increases over a period of 10-20 days. The growth rate of the cultures depends on the stage of growth of the cells when sub-cultured and the size of the initial inoculum. In rapidly growing cultures the doubling time is approximately 36-48 hrs and cell number may increase several hundred fold to a final density of approximately 3×10^6 cells/ml, but at low initial cell densities division occurs more slowly. Figure 2 shows that in complete contrast cell division does not occur when cultures are inoculated into a medium containing 2% lactose in place of sucrose (it should be noted that cell number is plotted on a linear scale in Figure 2).

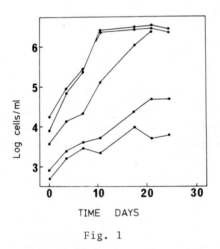

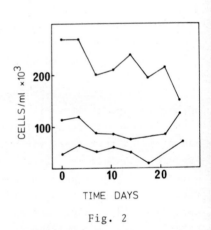

Fig. 1   Fig. 2

Fig. 1. The growth of sycamore cells in suspension culture. Syca-more cells were grown in 250 ml flasks on an orbital shaker at 25°C in a medium containing mineral salts, and 2% sucrose. Flasks were inoculated at different initial cell densities and growth monitored over a period of several weeks. Cell numbers were de-termined on a haemocytometer slide after treating the cells with 15% chromic acid for 5 mins at 70°C followed by mechanical shak-ing for 5 mins. Cell numbers are plotted on a log scale.

Fig. 2. The lack of growth of sycamore cells in lactose medium. Cells were inoculated into medium containing 2% lactose in place of sucrose. Other details were as described in the legend to Figure 1. Cell numbers are plotted on a linear scale.

PHAGE TREATMENT

In view of the fact that sycamore cells are unable to utilise lactose as a carbon source it was of interest to see whether they could
grow on lactose after pre-treatment with λ p *lac* 5 phage. Cells
grown in sucrose medium were treated for 2 days with the phage and the
cells were then washed several times and inoculated into fresh medium
containing lactose in place of sucrose.   In early experiments cell
growth was measured by the increase in volume of the cultures after
sedimentation of the cells for 2 hrs.   The results showed that although control cells did not grow on lactose those treated with λ *p*
lac 5 were capable of a substantial increase in cell volume (Johnson
et al., 1973).   In later experiments we measured directly the increase
in cell number in the cultures.   From the results shown in Figure 3
it is clear that although untreated cells and those treated with λ[+]
phage do not divide the λ p *lac* 5 treated cells may undergo several
rounds of cell division.   The rate of cell division is very low compared to cultures grown on sucrose and there is a certain amount of
variability in the timing of the response.   Generally it occurs between 4 and 8 weeks after placing the cells on lactose (compare Figures 3 and 4).   The cell divisions observed in the λ p *lac* 5 cells are
not sustained however and over a further period of several weeks the
cell number may actually fall.   Subculturing the cells into fresh
medium at intervals during incubation on lactose does not stimulate
division of the λ p *lac* 5 treated cells (Fig 4).   The failure of the
cells to continue growing on lactose is not due to the accumulation
of galactose in the medium (Fig 5).

FATE OF THE λ p *lac 5 PHAGE*.

When sycamore cells are treated with phage particles, the number of
plaque forming units that can be recovered from the culture steadily
declines over a 2-day period (Fig 6). Control experiments show that
this only occurs when intact sycamore cells are present and is not due
to inactivation of the phage by some component of the culture medium.
This suggests that the phage particles are either inactivated or taken
up by the sycamore cells and although the mechanism is not known the
results in Figure 6 suggest that this is a comparatively slow process.
For every cell present at the start of the 2-day incubation, approximately 10 phage particles are lost. At intervals during the incubation
period, samples of the treated cells were removed, washed and ground
with liquid nitrogen and aliquots of the extract tested for the presence of phage. No infective particles were recovered. One possible
explanation of this result is that the phage particles although pre-

sent are inactivated during the extraction or by the cell homogenate.
This possibility was examined by mixing infective phage with sycamore
cells and immediately breaking the cells by a number of different
methods. The results in Table 1 show that phage particles are not inactivated by these procedures.

Table 1
The survival of phage during
different cell extraction procedures.

	Number of phage recovered per ml	
	log phase cells	stationary phase cells
sand	8.0×10^7	2.5×10^7
liquid nitrogen	9.5×10^7	6.0×10^7
homogeniser	6.0×10^5	-
control	4.1×10^8	3.3×10^8

Half ml samples of λ p *lac* 5 phage were added to 5 ml
samples of log phase and stationary phase sycamore cells.
The suspensions were then ground by hand in a mortar and
pestle with acid washed sand or liquid nitrogen for five
mins, or homogenised at maximum speed with a Silveson homo-
geniser for five mins. Cell debris was removed by centrifugation and the extracts assayed for phage.

β-*GALACTOSIDASE ACTIVITY IN SYCAMORE CELLS*.

After treatment with phage λ p *lac* 5 carrying the Z gene for
β-galactosidase, sycamore cells are able to grow on lactose. One explanation of this result is that the genes for lactose utilisation
are taken up and expressed by the plant cells and one way to test this
hypothesis is to measure the β-galactosidase activity in the cells.
Figure 7 shows that untreated cells and those exposed to phage λ[+] and
λ p *lac* 5 all contain enzymes capable of hydrolysing O-nitrophenyl-
β-D-galactopyranoside, the substrate commonly used for β-galactosidase
assay. There is a transient increase in the enzyme activity of the
cells when they are transferred to lactose. This reaches a maximum
after about 8 days and the greatest activity is found in cells pretrea-
ted with phage λ p *lac* 5. It should be noted that this increase is
observed several weeks before the burst of cell division shown in
Figures 3 and 4. It seems clear from these results that β-galactosidase activity is not confined to those cells able to grow on lactose.

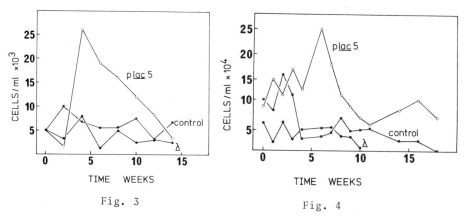

Fig. 3   Fig. 4

Fig. 3. The growth of phage-treated cells in lactose. Phage λ^+ and
λ p *lac* 5 were grown up on the *E. coli lac* deletion strain ED914,
harvested and passed through a 0.45 μ nitrocellulose filter.
Sycamore cells (1.8 x 10^5/ml) grown in sucrose were treated with
either λ^+ or λ p *lac* phage at a final concentration of 10^9 p.f.u./
ml for 2 days. The cells were then washed in and transferred to
lactose medium and changes in cell number measured over a period
of several weeks. Cell numbers are plotted on a linear scale.

Fig. 4. The subculturing of phage-treated cells grown in lactose medium. Cells were treated as described in the legend to Figure 3
except that they were transferred to fresh lactose medium after
2, 4, 6, 8 and 10 weeks.

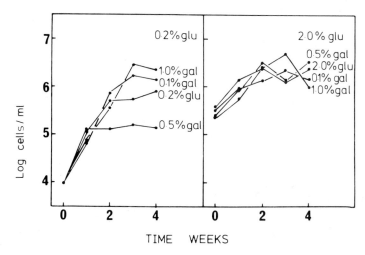

Fig. 5. The effect of galactose on the growth of sycamore cells in glucose medium. Cells were grown in medium containing 2% or 0.2% glucose in place of sucrose plus varying concentrations of galactose.
Cell numbers are plotted on a log scale.

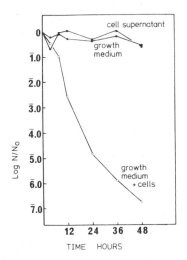

Fig. 6. The disappearance of λ p *lac* 5 phage during incubation with
sycamore cells in culture. Five ml aliquots of λ p *lac* 5 phage
in nutrient broth was added to 15 ml of sycamore cells in sucrose
medium giving a final concentration of 5×10^5 sycamore cells
and 10^9 p.f.u./ml. A similar concentration of phage was added
to fresh medium and to the supernatant from a stationary phase
cell culture. At intervals over a 2-day period the number of
phage particles in the medium was determined by the agar layer
plating method. Samples of the sycamore cells were withdrawn
after 0, 1 and 2 days, washed, ground in liquid nitrogen, resuspended in phage buffer and assayed for phage. No phage were
detected in the cell extracts.

Fig. 7. β-galactosidase activity in control and phage-treated cells
grown in lactose medium. Sycamore cell cultures were treated for
2 days with phage λ+ or λ p *lac* 5 and the cells transferred to
lactose medium. At intervals samples of the cells were filtered
and ground after freezing with liquid nitrogen. One ml of 0.1 M
phosphate buffer pH 7.2 was added and β-galactosidase activity
measured with O-nitrophenyl β-D-galactopyranoside as substrate
(Colowick and Kaplan, 1955).

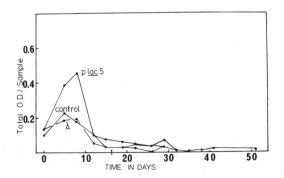

Animal cells are known to contain enzymes with specificity towards
β-galactosides and recently it has been shown that there are two β-gal-
actosidases in E. coli K12 (Hartle and Hall, 1974). Strains in which
the Z gene is deleted are unable to grow on lactose although they contain β-galactosidase II. This enzyme is induced by lactose but has
very little lactase activity. A similar situation may exist in sycamore cells for the β-galactosidase detected here is induced by lactose.
The failure of the cells to grow suggests that this enzyme does not
metabolise lactose. The question remains as to whether any enzyme
coded for by the Z gene is present in the λ p lac 5 treated cells.
Our approach to this has been to test the cell extracts with an antibody to the E. coli β-galactosidase but at present there is no firm
evidence for the synthesis of the bacterial enzyme.

DISCUSSION

The aim of these experiments is to investigate whether plant cells
can take up phage particles and utilise the genetic information, and
if this is the case to establish whether this information can be incorporated into the cells in a stable manner. The sycamore cells do not
normally grow on lactose. This provides a favourable situation for
selecting cells that may have acquired the genes for lactose utilisation from the lac transducing phage.

The criterion of growth on lactose is not sufficient evidence in
itself to conclude that uptake and expression of the lac genes is taking place, particularly as the treated cells grow only very slowly and
growth is only a transient phenomenon. The results shown in Figure 6
suggest that intact sycamore cells render the phage incapable of infecting E. coli. This could occur for example if the phage particle
was disrupted or more simply if the adsorption site was made inopera-
tive. It is not known whether the phage particles enter the plant
cells during this process although the use of radioactive phage labelled in the DNA and the protein coat should answer this question. No
factors responsible for phage inactivation can be detected in the culture medium or in cell homogenates.

It does seem possible for plant cells to take up phage particles and
express the genetic information. Experiments by Carlson (1973) have
shown that when protoplasts of Hordeum vulgare are exposed to phage
T3 there is a rapid synthesis of phage-specific RNA polymerase and the
s-adenosylmethionine-cleaving enzyme. In our experiments there is no
conclusive evidence that the E. coli β-galactosidase is synthesised in
the sycamore cells. This question is being investigated by the use of

antibodies. Experiments conceptually similar to ours have also been carried out by Doy and his colleagues (Doy *et al.*, 1973). They have used haploid callus cultures of *Lycopersicon esculentum* and *Arabidopsis thaliana* treated with phage ϕ 80 p *lac*$^+$ and λ p *gal*$^+$. In these experiments they claim to have shown the synthesis of the *E. coli* β-galactosidase using an antibody heat-protection assay for the enzyme.

We have tried a number of approaches to improve the growth response of the λ p *lac* 5-treated sycamore cells on lactose. Typically the results show that a burst or cell division occurs but that this is not maintained. Subculturing the cells does not stimulate further division. Control experiments show that the failure of the cells to continue growing is not due to the build-up of galactose in the culture medium. It is possible that the phage DNA is expressed in the plant cells but that it is eventually inactivated or destroyed. An alternative explanation is that expression of the phage DNA initially allows growth to take place on lactose but that the synthesis of other phage gene products that are deleterious to the cells also occurs. This would explain why the cell number actually falls after the initial cell divisions.

ACKNOWLEDGEMENT

This work was supported by a grant from the Science Research Council.

REFERENCES

Carlson, P.S. (1973). The use of protoplasts for genetic research. *Proceedings of the National Academy of Sciences, U.S.A.* 70, 598-602.

Colowick, S.P., and Kaplan, N.O. (1955). β-galactosidase (lactose) from *E. coli*. Assay method 1. *Methods in Enzymology*. 1, 241. Academic Press.

Doy, C.H., Gresshof, P.M., and Rolfe, B.G. (1973). Biological and molecular evidence for the transgenosis of genes from bacteria to plant cells. *Proceedings of the National Academy of Sciences, U.S.A.* 70, 723-726.

Hartl, D.L., and Hall, B. (1974). Second naturally occurring β-galactosidase in *E. coli*. *Nature*. 248, 152-153.

Ippen, K., Shapiro, J.A., and Beckwith, J.R. (1971). Transposition of the *lac* region to the *gal* region of the *E. coli* chromosome: isolation of λ *lac* transducing bacteriophages. *Journal of Bacteriology*. 108, 5-9.

Johnson, C.B., and Grierson, D. (1974). The uptake and expression of DNA by plants. In *Current Advances in Plant Science,* Editor H.Smith, Volume 2, No. 9, 1-12.

Johnson, C.B., Grierson, D., and Smith, H. (1973). Expression of λ p
   lac 5 DNA in cultured cells of a higher plant. *Nature New Biology*.
   244, 105-107.
Ledoux, L., Huart, R., and Jacobs, M. (1974). DNA-mediated genetic correction of thiamineless *Arabidopsis thaliana*. *Nature*. 249, 17-21.

REPLICATION AND EXPRESSION OF TOBACCO MOSAIC VIRUS GENOME
IN ISOLATED TOBACCO LEAF PROTOPLASTS

ITARU TAKEBE, SHIGEJI AOKI[*] AND FUKUMI SAKAI

Institute for Plant Virus Research,
959 Aobacho, Chiba 280, Japan.

Synthesis of tobacco mosaic virus (TMV)-specific
RNAs and proteins was studied using isolated tobacco
leaf protoplasts inoculated *in vitro*. Double-stranded
replicative form and replicative intermediate of TMV-RNA
were identified in infected protoplasts, and their
involvement in the replication of viral RNA was
suggested. The time course of viral RNA synthesis
showed that progeny viral RNA accumulates in free or
partially coated forms during early periods of infection, whereas in later periods most of TMV-RNA are
incorporated into virus particles. In addition to
viral coat protein, two high molecular weight non-coat
proteins were synthesized in infected protoplasts.
Maximal synthesis of the latter proteins occurred
several hours before that of coat protein. The mode
of *in vivo* replication and translation of TMV-RNA is
discussed in comparison with that of bacteriophage RNA.

Recent progress in the isolation of protoplasts from higher plant
tissues (Cocking, 1972) has stimulated the use of this wall-less form
of plant cells in a variety of biological disciplines. In particular,
protoplasts isolated from tobacco leaves have been used most widely
and have proved to be an extremely valuable material for plant virology (Takebe, 1975) which has long suffered from a lack of suitable
experimental systems for the detailed analysis of infection processes.
In this article we shall summarize the knowledge on the synthesis and
the translation of tobacco mosaic virus (TMV) RNA in isolated tobacco
leaf protoplasts. Studies of the replication and the expression of
a viral genome should not only be of central importance for plant

---

[*] Present address: Department of Microbiology, Nippon Dental College,
1-8 Hamauracho, Niigata 951, Japan.

virology but also provide a basis for understanding how foreign genetic material might be proliferated and expressed in plant cells.

DESCRIPTION OF THE SYSTEM.

Protoplasts of palisade parenchyma cells are prepared in large
amounts (up to 10^7 protoplasts/ g fresh weight of leaves) by sequen-
tially digesting mature tobacco leaves with polygalacturonase and
cellulase under plasmolyzing conditions (Takebe *et al.*, 1968; Otsuki
et al., 1974). The isolated tobacco leaf protoplasts survive as such
in a simple medium consisting of salts and plant hormones (Takebe *et
al.*, 1968), whereas they form new surface walls, undergo cell division, and develop into callus-like aggregates of progeny cells in
richer media (Nagata and Takebe, 1970, 1971; Takebe *et al.*, 1971).
Inoculation with TMV is performed by adding purified virus to the
protoplast suspension and by incubating the mixture for 10 min
(Takebe and Otsuki, 1969; Otsuki *et al.*, 1972). Presence of a polycation such as poly-L-ornithine in the inoculum is essential for infection to occur (Takebe and Otsuki, 1969), and it is believed that
the polycation helps the virus to adsorb to and penetrate the plasmalemma (Otsuki *et al.*, 1972). Over 90% of the total number of protoplasts are infected under optimal conditions as examined by staining
with fluorescence-labelled TMV antibody (Otsuki and Takebe, 1969)
(Fig 1). Infected protoplasts produce more than 10^6 TMV particles per
cell within 2-3 days.

The most important feature of the infection in protoplasts is that
it occurs synchronously in nearly all the cells. Since poly-L-ornithine is necessary for the infection of protoplasts, there is little
chance that progeny virus causes secondary infection. The system,
therefore, permits studies of a type which has not been feasible with
conventional materials, namely, to follow the events occurring during
a single cycle of infection in individual cells. Furthermore, the
protoplast system is particularly suitable for quantitative biochemical experimentation, because it forms a uniform suspension and allows
ready extraction of intracellular substances.

SYNTHESIS OF TMV-*SPECIFIC* RNAs *(AOKI AND TAKEBE, 1975).*

Electrophoretic separation of RNAs in polyacrylamide gels (Loening,
1967) provides a convenient and accurate means of analysing virus-
specific RNAs synthesized in infected protoplasts. RNA synthesis in
uninfected protoplasts can be effectively inhibited by actinomycin D

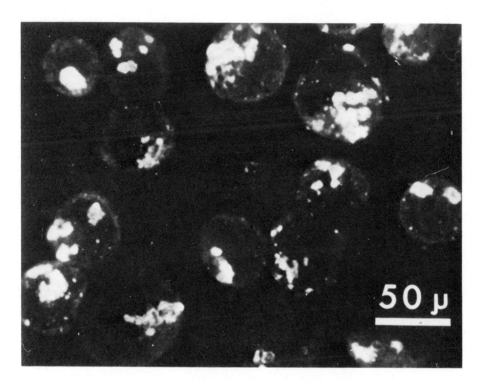

Fig. 1. Tobacco leaf protoplasts inoculated with tobacco mosaic virus
and stained after 24 hrs of culture with fluorescence-
labelled viral antibody.

and only small amounts of ^{32}P-radioactivity are incorporated into 25S
and 18S cytoplasmic ribosomal RNAs (Fig 2) as well as into transfer
RNA. Infection by TMV results in the synthesis of three additional
RNA species in protoplasts (Fig 2B), and these were identified as
single-stranded viral RNA, its replicative form (RF) and replicative
intermediate (RI).

The fastest moving species of virus-specific RNAs (slices 16-23
in Fig 2B) has the same mobility as TMV-RNA and is eluted from a
cellulose column with a buffer containing 15% ethanol (Fig 3B). In
contrast, the two slower moving species are eluted with butter containing no ethanol (Fig 3C), indicating their double-stranded structures. The faster moving component of TMV-specific double-stranded
RNAs (slices 15-17 in Fig 2B) is soluble in M NaCl solution (Fig 3D)
and is resistant to RNAase. In contrast, the slower moving component
(slices 1-7 in Fig 2B) is insoluble in M NaCl (Fig 3E) and is

converted into the faster moving component by RNAase digestion (Fig 3F). These results show that the faster moving component of TMV-specific double-stranded RNAs has a purely double-stranded structure of uniform size, whereas the slower moving component has the same basic structure to which single-stranded tails are attached.

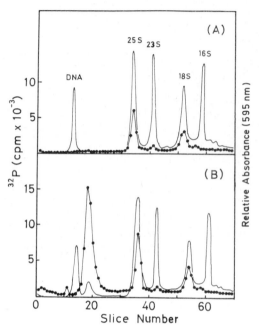

Fig. 2. Gel electrophoresis profile of nucleic acids extracted from uninfected (A) and infected (B) protoplasts. Gels were stained with methylene blue. Absorbance at 595 nm (———) and radioactivity (–•———•–).

Denaturation with dimethylsulphoxide of the faster moving component of TMV-specific double-stranded RNAs yields an RNA with the same size as TMV-RNA as the main product (Fig 4). The molecular weight of this component was $3.7-4.0 \times 10^6$ Daltons, as determined by coelectrophoresis with double-stranded rice dwarf virus RNA. Furthermore, its nucleotide composition was shown to closely agree with that expected for a structure consisting of TMV-RNA and its complementary strand (Table 1). All these results are consistent with the idea that the faster moving component of TMV-specific double-stranded RNAs is RF of TMV-RNA. The slower moving component should then be RI, because it is converted into RF by RNAase treatment (Fig 3F) and yields on denaturation an RNA

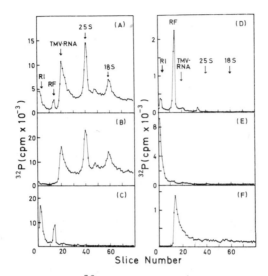

Fig. 3. Fractionation of ^{32}P-labelled RNA extracted from TMV-infected protoplasts. (A) unfractionated RNA. (B) RNA eluted from a cellulose column with 15% ethanol, (C) RNA eluted from the same column with buffer containing no ethanol, (D) M NaCl soluble RNA of C, (E) M NaCl insoluble RNA of C, (F) E after RNAase digestion.

with the size of TMV-RNA together with heterogeneous population of smaller RNA species.

It has been reported that some other forms of TMV-specific RNA occur in infected leaf tissues (Jackson et al., 1972) and one of these was identified as a fragment of TMV-RNA (Siegel et al., 1973). These forms were not found in infected protoplasts.

The time course of synthesis of TMV-specific RNAs was studied by culturing inoculated protoplasts in the presence of ^{32}P-phosphate and by analyzing their RNA at various times post infection by gel electrophoresis. The amount of single-stranded viral RNA could be determined by measuring both the radioactivity and the absorbance of corresponding gel slices, thus permitting its specific radioactivity to be calculated. This was used in turn to calculate the absolute amount of RF and RI from their radioactivity. A parallel culture was run to follow the production of virus particles by assaying the infectivity in protoplast extracts.

106

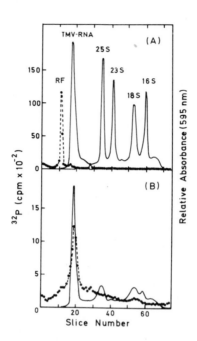

Fig. 4. Gel electrophoresis of ^{32}P-labelled RF before (A) and after (B) denaturation with dimethylsulphoxide. Absorbance at 595 nm (———) and radioactivity (—●———●—).

Table 1

Nucleotide composition of ^{32}P-labelled RF prepared from TMV-infected protoplasts.

Nucleo-tide	TMV-RNA*	RF**	Mole percent ^{32}P-RF from protoplasts			
			Exp. 1	Exp. 2	Exp. 3	Exp. 4
AMP	29.8	28.0	28.4	28.3	28.1	28.5
GMP	25.3	22.0	21.5	21.2	21.6	21.4
CMP	18.5	22.0	21.7	21.2	22.3	20.9
UMP	26.3	28.0	28.4	29.3	28.0	29.2

* From Knight (1952).
** Theoretical values calculated from the nucleotide composition of TMV-RNA.

As shown in Figure 5, synthesis of viral RNA was detectable 4 hrs post infection and proceeded initially at an exponential rate. Viral RNA synthesis probably started earlier, since more than 10^3 TMV-RNA molecules were present per protoplast at this time. Vigorous synthesis of viral RNA thus preceded that of virus particles by 4-5 hrs. The rate of viral RNA synthesis levelled off rather sharply at 8 hrs post infection, at which time virus particles were being produced at the maximal rate. After this time viral RNA was synthesized at a low rate, and the curves for viral RNA synthesis and particle production practically overlapped in later periods of infection. The time course of synthesis of RI virtually paralleled that of viral RNA synthesis, whereas RF did not increase in the later periods. It is clear from the results in Figure 5 that free or only partially coated viral RNA molecules accumulate in protoplasts during the early period of infection. Most of viral RNA appears to exist as complete virus particles during the later periods.

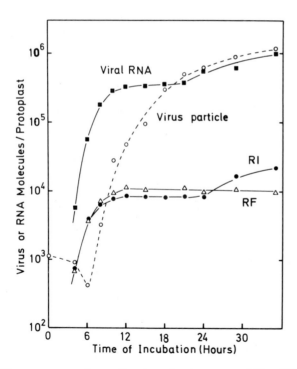

Fig. 5. Time course of synthesis of virus-specific RNAs and virus particles in TMV-inoculated tobacco leaf protoplasts.

The time course of synthesis of virus-specific RNAs suggested that the double-stranded RNA forms, in particular RI, are involved in the replication of TMV-RNA. In pulse-chase type experiments, ^{3}H-uridine radioactivity incorporated into RF and RI during 30 min of labelling decreased by chase, whereas it increased steadily during continuous labelling. The radioactivity in both RF and RI could not be chased out completely, however, probably reflecting the large pool size of RNA precursors. Using leaf cells isolated from TMV-infected tobacco, it has been reported that the radioactivity in RI but not in RF is completely chased out (Jackson et al., 1972). Although the pulse-chase experiments also suggest the intermediate role of double-stranded RNAs for TMV-RNA replication, conclusive demonstration of their role should await more detailed studies, including those with isolated replicase.

SYNTHESIS OF TMV-SPECIFIC PROTEINS (SAKAI AND TAKEBE, 1972, 1974).

TMV-RNA having a molecule weight of 2×10^6 Daltons is large enough for potentially coding several proteins including replicase. However, viral coat protein is the only protein whose information has been shown to reside in TMV-RNA (Wittmann, 1962). We have been trying to identify translation products of TMV-RNA in infected tobacco leaf protoplasts.

A major technical difficulty encountered in such efforts arises from the fact that the amount of proteins in question is extremely small, at least in the early stages of infection. Since host protein synthesis is neither affected by infection nor can be inhibited efficiently with actinomycin D (Sakai and Takebe, 1970), virus-specific proteins have to be detected among overwhelming amounts of host proteins. We found that this difficulty can be largely eliminated by irradiating inoculated protoplasts with ultraviolet (UV) light. Both RNA and protein synthesis in uninfected protoplasts is severely inhibited by low UV doses (Fig 6), whereas TMV multiplication is relatively insensitive to UV irradiation. It is, therefore, possible with appropriate UV doses to reduce the rate of host protein synthesis down to a level comparable to that of synthesis of virus-specific proteins (Fig 7).

Fig. 6. Effects of ultraviolet irradiation on RNA and protein synthesis in uninfected protoplasts. Syntheses of RNA and protein were measured by incorporation of ^{14}C-uracil and ^{14}C-leucine, respectively. Intensity of ultraviolet light was 38 erg mm^{-2} sec^{-1}.

Proteins synthesized in TMV-inoculated and irradiated protoplasts were analyzed by electrophoresis in sodium dodecylsulphate (SDS)-polyacrylamide gels. Three species of virus-specific proteins were detected in protoplasts labelled with ^{14}C-leucine (Fig 8). The largest and the fastest moving peak represents viral coat protein monomer, since it has the same mobility as an authentic sample and since it is not labelled with radioactive histidine or methionine, amino acids absent in the coat protein. The other two TMV-specific proteins are characterized by high molecular weights (140,000 and 180,000 Daltons) and most probably are unrelated to coat protein, since they are also labelled with histidine and methionine. These proteins apparently correspond to the 155,000 and 195,000 Dalton proteins which have been detected in infected tobacco leaves (Zaitlin and Hariharasubramanian, 1972). It should be emphasized that coat protein is synthesized in far greater amounts than the high molecular weight proteins (Fig 8). The difference in terms of the number of molecules synthesized is probably greater than is judged from the amount of radioactivity, because coat protein has a low molecular weight.

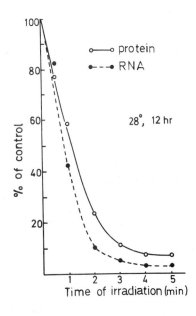

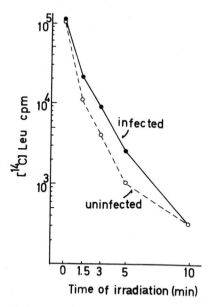

Fig. 7. Effects of ultraviolet irradiation on protein synthesis in infected and uninfected protoplasts.

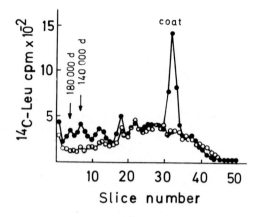

Fig. 8. SDS-gel electrophoresis profiles of proteins extracted from infected (●————●) and uninfected (o————o) protoplasts.

The TMV-specific protein with a molecular weight of 140,000 Daltons was studied in some detail. It does not contain the coat protein sequence, and consists of a single large polypeptide, since its electrophoretic mobility is not altered by treatment with 6 M guanidine (Fig 9), 1 M urea or heat. This protein is present in a subcellular fraction sedimenting between 1,000 and 20,000 x g, and appears, therefore, to be located in some cellular structure lighter than nuclei or chloroplasts. The same subcellular fraction is known to contain an enzyme which catalyzes the synthesis of double-stranded forms of TMV-RNA (Bradley and Zaitlin, 1971). This enzyme was recently solubilized and was shown to have a molecular weight approximating 140,000 Daltons (Zaitlin et al., 1973). These findings strongly suggest that the 140,000 Dalton protein synthesized in protoplasts is the replicase of TMV-RNA. It has been reported that frog oocytes translate injected

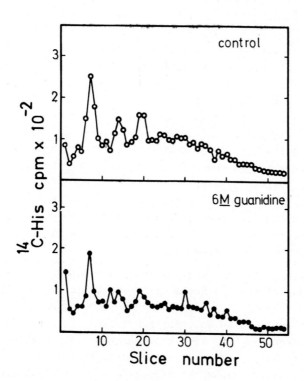

Fig. 9. Effect of denaturation with 6 M guanidine on the electrophoretic mobility of the TMV-specific 140,000 Dalton protein. Protein was extracted from the 1,000-20,000 x g fraction of infected protoplasts.

TMV-RNA and produce a protein with a molecular weight of 140,000
Daltons (Knowland, 1973). This protein appears to have some properties in common with the 140,000 Dalton protein in tobacco protoplasts.
Taking the available information together, it is highly probable that
the 140,000 Dalton protein synthesized in TMV-infected protoplasts is
coded by TMV-RNA. Approximately 60% of the potential coding capacity
of TMV-RNA is then accounted for by this protein.

Little is known about the nature and the function of 180,000 Dalton
protein. It is clear that TMV-RNA cannot accommodate the cistrons of
both the 140,000 and 180,000 Dalton proteins, unless they have a common sequence.

The time course of the synthesis of the three TMV-specific proteins
was studied by labelling TMV-inoculated irradiated protoplasts with
^{14}C-leucine. The same protoplasts were also labelled with ^{3}H-uridine
to follow the replication of viral RNA and to correlate it with the
synthesis of TMV-specific proteins. Figure 10 shows the temporal
relationships of the synthesis of the viral RNA, coat and the 140,000
Dalton protein as well as the virus particles. It may be clearly
seen that the two TMV-specific proteins follow different time courses
of synthesis. The 140,000 Dalton protein attained a maximal rate of
synthesis 4 hrs earlier than did the coat protein, but the rate de-
clined in later stages of infection. The course of synthesis of this
protein thus nearly paralleled that of viral RNA. In contrast, coat
protein synthesis lagged behind that of viral RNA, but continued at
a near maximal rate throughout the later periods. Synthesis of coat
protein was closely followed by the production of complete virus
particles. Although the course of synthesis of 180,000 Dalton protein
could not be traced accurately, it appeared to be rather similar to
that of 140,000 Dalton protein.

REGULATION OF VIRAL RNA REPLICATION AND TRANSLATION.

The knowledge of the synthesis of TMV-specific RNAs and proteins
in protoplasts is obviously too rudimentary to depict a detailed
picture of the replication and the translation of viral RNA in infected cells. However, the available information may allow us to
make a tentative sketch of what is occurring in individual infected
cells, with some ideas about the regulation of viral RNA replication
and translation. The following considerations are based on the
assumption that the 140,000 Dalton protein is a translation product
of viral RNA and represents the replicase of TMV-RNA.

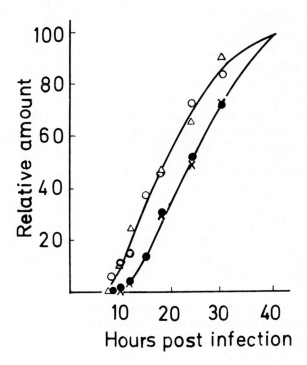

Fig. 10. Time course of synthesis of viral RNA (Δ), 140,000 Dalton
protein (o), coat protein (●) and virus particles (×)
in TMV-inoculated ultraviolet-irradiated protoplasts.
Data are expressed as percentage of the maximum values.

It may reasonably be assumed that the replicase cistron of parental RNA is translated when the inoculum virus enters the protoplasts
and is uncoated. The first replicase molecules thus synthesized use
the parental RNA as a template and produce progeny RNA molecules via
double-stranded intermediate forms. The progeny RNA molecules then
serve as a messenger for more replicase to be synthesized on one
hand, and as a template for replication of viral RNA on the other
hand. Both viral RNA and replicase are thus synthesized with an exponential rate during the initial period of infection. It should be
noted that these events can take place in a cell at more than one site,
since two strains of TMV have been shown to be able to establish infection in individual protoplasts (Otsuki and Takebe, unpublished).

Most of the viral RNA synthesized during early periods is not
immediately incorporated into virus rods but remains either free or

in partially coated forms. This is because the production of coat protein cannot keep pace with the exponential synthesis of viral RNA, especially during the initial period when the amount of available messenger is very small. Studies on the *in vitro* assembly of TMV show that the formation of a disk consisting of 34 coat protein sub-units is the first step of viral rod formation (Okada and Ohno, 1972). Coat protein must, therefore, accumulate to a certain level before massive formation of virus particles can proceed. It should also be recalled that as many as 2130 coat protein molecules are required to encapsidate each molecule of TMV-RNA.

The rate of coat protein synthesis rapidly rises as free progeny RNA molecules accumulate, and the production of large amounts of coat protein now lead to the encapsidation of a substantial fraction of progeny RNA. This will in turn cause the shortage not only of the replicase messenger but also of the template for viral RNA replication, and should consequently slow down the synthesis of both replicase and viral RNA. It is also possible that coat protein binds more specifi-cally to the replicase cistron of viral RNA and prevents its trans-lation, while allowing a continued production of coat protein. A similar mechanism for the regulation of replicase production has been known for RNA bacteriophages (Kozak and Nathans, 1972).

It is possible that the rapid coat protein production and the slow viral RNA synthesis during the intermediate periods of infection give rise to an oversupply of coat protein, so that any viral RNA newly synthesized in later periods is rapidly encapsidated. Viral RNA synthesis during the later periods of infection probably utilizes preexisting replicase and the rate is limited by the availability of viral RNA as the template for replication.

The mode of *in vivo* replication and translation of TMV-RNA emer-ging from the studies with protoplasts appears to be similar in princi-ple to that of bacteriophage RNA. Particularly noteworthy are the observations that the coat protein cistron is translated much more frequently than the cistron for the 140,000 Dalton protein, and that the synthesis of the two virus-specific proteins becomes maximal at different times after infection. These features are analogous to those known for the proteins of RNA bacteriophages (Kozak and Nathans, 1972) and indicate clearly that individual cistrons of TMV-RNA are translated separately. It is difficult to reconcile these results with the monocistronic mode of translation of mammalian viral RNA by which the entire span of viral RNA is translated into a large poly-

peptide which is later split into individual virus-specific proteins
(Jacobson and Baltimore, 1968; Summers and Maizel, 1968). A polycistronic model for the translation of TMV-RNA has been proposed on
the basis of the size of TMV-specific polysomes (Kiho, 1972). Occurrence of plant viruses with divided genomes may have relevance to
this type of viral RNA translation.

It is perhaps appropriate to point out, however, that there are
many indications of differences between plant and bacterial RNA viruses with respect to details in the replication and the translation
of their RNA. For example, plant viral RNA has twice as much information content as bacteriophage RNA, yet there has been no evidence
for the presence in plant viruses of a protein which is analogous to
the bacteriophage maturation protein. It is, therefore, quite possible that the number and the order of cistrons in plant viral genomes
are different from those of RNA bacteriophages. Furthermore, the
available evidence suggests that the replicase of TMV is a large
single polypeptide, whereas the Qβ replicase consists of four subunits
only one of which is coded by the viral genome (Kondo et al., 1970).
Finally, the TMV-replicase is apparently associated with cellular
structures, whereas the bacteriophage replicase is found in the soluble fraction. Double-stranded forms of plant viral RNA also occur
predominantly in a particulate fraction (Ralph and Wojcik, 1969).
These observations strongly indicate that the replication of plant
viral RNA takes place at some subcellular structure. It should be
evident that structural organization of plant cells have to be taken
into account to fully understand the in vivo replication and translation of plant viral RNAs, and their regulation.

ACKNOWLEDGEMENT

This work was partially supported by a project grant from the
Ministry of Agriculture and Forestry.

REFERENCES

Aoki, S., and Takebe, I. (1975). Replication of tobacco mosaic virus
    RNA in tobacco mesophyll protoplasts inoculated in vitro. Sub-
    mitted for publication.
Bradley, D.W., and Zaitlin, M. (1971). Replication of tobacco mosaic
    virus. II. The in vitro synthesis of high molecular weight virus-
    specific RNAs. Virology. 45, 192-199.
Cocking, E.C. (1972). Plant cell protoplasts--isolation and develop-
    ment. Annual Review of Plant Physiology. 23, 29-50.

Jackson, A.O., Zaitlin, M., Siegel, A., and Francki, R.I.B. (1972). Replication of tobacco mosaic virus. III. Viral RNA metabolism in separated leaf cells. *Virology*. 48, 655-665.

Jacobson, M.F., and Baltimore, D. (1968). Polypeptide cleavages in the formation of poliovirus proteins. *Proceedings of the National Academy of Sciences, U.S.A.* 61, 77-84.

Kiho, Y. (1972). Polycistronic translation of plant viral ribonucleic acid. *Japanese Journal of Microbiology*. 16, 259-267.

Knight, C.A. (1952). The nucleic acids of some strains of tobacco mosaic virus. *Journal of Biological Chemistry*. 197, 241-249.

Knowland, J. (1973). Protein synthesis directed by the RNA from a plant virus in a normal animal cell. *Genetics*, in press.

Kondo, M., Gallerani, R., and Weissmann, C. (1970). Subunit structure of Qβ replicase. *Nature*. 228, 525-527.

Kozak, M., and Nathans, D. (1972). Translation of the genome of a ribonucleic acid bacteriophage. *Bacteriological Revue*. 36, 109-134.

Loening, U.E. (1967). The fractionation of high-molecular weight ribonucleic acid by polyacrylamide-gel electrophoresis. *Biochemical Journal*. 102, 251-257.

Nagata, T., and Takebe, I. (1970). Cell wall regeneration and cell division in isolated tobacco mesophyll protoplasts. *Planta*. 92, 301-308.

Nagata, T., and Takebe, I. (1971). Plating of isolated tobacco mesophyll protoplasts on agar medium. *Planta*. 99, 12-20.

Okada, Y., and Ohno, T. (1972). Assembly mechanism of tobacco mosaic virus particle from its ribonucleic acid and protein. *Molecular General Genetics*. 114, 205-213.

Otsuki, Y., and Takebe, I. (1969). Fluorescent antibody staining of tobacco mosaic virus antigen in tobacco mesophyll protoplasts. *Virology*. 38, 497-499.

Otsuki, Y., Takebe, I., Matsui, C., and Honda, Y. (1972). Ultrastructure of infection of tobacco mesophyll protoplasts by tobacco mosaic virus. *Virology*. 49, 188-194.

Otsuki, Y., Takebe, I., Honda, Y., Kajita, S., and Matsui, C. (1974). Infection of tobacco mesophyll protoplasts by potato virus X. *Journal of General Virology*. 22, 375-385.

Ralph, R.K., and Wojcik, S.J. (1969). Double-stranded tobacco mosaic virus RNA. *Virology*. 37, 276-282.

Sakai, F., and Takebe, I. (1970). RNA and protein synthesis in protoplasts isolated from tobacco leaves. *Biochimica et Biophysica Acta*. 224, 531-540.

Sakai, F., and Takebe, I. (1972). A non-coat protein synthesized in tobacco mesophyll protoplasts infected by tobacco mosaic virus. *Molecular General Genetics*. 118, 93-96.

Sakai, F., and Takebe, I. (1974). Protein synthesis induced in tobacco mesophyll protoplasts by tobacco mosaic virus infection. *Virology*, in press.

Siegel, A., Zaitlin, M., and Duda, C.T. (1973). Replication of tobacco mosaic virus. IV. Further characterization of viral related RNAs. *Virology*. 53, 75-83.

Summers, D.F., and Maizel, J.V. Jr. (1968). Evidence for large precursor proteins in poliovirus synthesis. *Proceedings of the National Academy of Sciences, U.S.A.* 59, 966-971.

Takebe, I., Otsuki, Y., and Aoki, S. (1968). Isolation of tobacco mesophyll cells in intact and active state. *Plant and Cell Physiology*. 9, 115-124.

Takebe, I., and Otsuki, Y. (1969). Infection of tobacco mesophyll protoplasts by tobacco mosaic virus. *Proceedings of the National Academy of Sciences, U.S.A.* 64, 843-848.

Takebe, I., Labib, G., and Melchers, G. (1971). Regeneration of whole plants from isolated mesophyll protoplasts of tobacco. *Naturwissenschaften*. 58, 318-320.

Takebe, I., (1975). The use of protoplasts in plant virology. *Annual Review of Phytopathology*. 13, in press.

Wittmann, H.G. (1962). Proteinuntersuchungen an Mutanten des Tabakmosaikvirus als Beitrag zum Problem des genetischen Codes. *Zeitschrift für Vererbungslehre*. 93, 491-530.

Zaitlin, M., and Hariharasubramanian, V. (1972). A gel electrophoretic analysis of proteins from plants infected with tobacco mosaic and potato spindle tuber viruses. *Virology*. 47, 296-305.

Zaitlin, M., Duda, C.T., and Petti, M.A. (1973). Replication of tobacco mosaic virus. V. Properties of the bound and solubilized replicase. *Virology*. 53, 300-311.

PLANT PROTOPLASTS IN TRANSFORMATION STUDIES:
SOME PRACTICAL CONSIDERATIONS

J.W. WATTS, D. COOPER AND JANET M. KING

John Innes Institute,
Colney Lane,
Norwich.

The infection of tobacco leaf protoplasts with viral
RNA has been used to determine the optimal conditions
for introducing nucleic acids into plant protoplasts.
A polycation (e.g. poly-L-ornithine) was essential and
the condition of the protoplasts was important. Relatively large amounts of RNA entered the protoplasts
within the first minute of inoculation. The procedure
has been used to inoculate tobacco protoplasts with
DNA of *Agrobacterium tumefaciens*. No auxin-independent
cells were produced.

The search for new and improved methods in plant breeding has
directed attention to a number of promising approaches, for example,
the use of haploids (Melchers, 1972), hybridization through somatic
cells (Tempé, 1973), and transformation of cells with DNA (Ledoux,
1971). Several groups are currently investigating transformation of
plant material with some indications of success. Transformation has
been claimed with plant DNA on petunia (Hess, 1972), bacterial DNA
on *Arabidopsis* (Ledoux, Huart and Jacobs, 1973) and bacteriophage DNA
on plant cells in tissue culture (Doy, Gresshoff and Rolfe, 1973;
Johnson, Grierson and Smith, 1973). The technique is at present however little more than a theoretical possibility because the practical
problems have hardly begun to be defined. The results already obtained are in most cases extremely confusing and the nature of the process
has not been elucidated.

The process of transformation can be divided somewhat arbitrarily
into four stages: the preparation of suitable DNA, introduction of
the DNA into the cell, integration of DNA with the genome, and finally,
expression. We may perhaps have some confidence that suitable preparations of DNA can be prepared, but the later stages are not understood, so it is by no means clear whether they can be achieved. One
of the problems that has confronted all efforts to introduce DNA into
plant cells is the barrier of the cell wall. Very definite claims

for uptake of DNA by cells have been made by workers from several laboratories, but there is considerable disagreement about the state and condition of the DNA or its degradation products in the cell (see Ledoux, 1971 for discussions). As a result, some attention is being given to the use of protoplasts in transformation work (Ohyama, Gamborg and Miller, 1972). What is lacking however in all the work on transformation is a sensitive assay to determine whether DNA enters the cell in a competent state. The ultimate criterion is that transformation should occur, but it would be useful to have some perhaps less demanding measure of the early stages. The protoplast-virus system offers a method for determining the conditions necessary for introducing nucleic acids into cells. Ideally, a plant virus containing DNA (cauliflower mosaic virus) would be preferred, but for technical reasons this is not at present feasible. The viruses containing RNA are however suitable for studying the problem of entrance into the protoplast, and the development of virus within protoplasts is a stringent test of the quality of virus or virus RNA that has entered it. Cowpea chlorotic mottle virus (CCMV) is a small spherical virus which possesses a divided genome; infection requires three different particles containing four species of RNA of which three are essential for infection (Bancroft and Flack, 1972). Infection therefore requires that on average a minimum of about 10 particles should enter the cell. CCMV and its RNA infect tobacco protoplasts (Motoyoshi et al., 1973a) whilst regeneration of whole plants from tobacco protoplasts is an established procedure (Nagata and Takebe, 1971). We have therefore used this system to establish the conditions required for uptake of nucleic acids into protoplasts and have also attempted to use the DNA of *Agrobacterium tumefaciens* to promote auxin independence in tobacco protoplasts.

METHODS

The protoplasts used in this work were prepared from leaves of *Nicotiana tabacum* (cv. White Burley) by the method of Motoyoshi et al. (1973a). When aseptic conditions were used the leaves were sterilized for 20 mins in 1% sodium hypochlorite solution. Half-expanded leaves were taken from plants 40-45 days old; the lower epidermis was removed and 2 g portions of leaf were treated with successive 20 ml volumes of 0.5% Macerozyme R10, 0.5% dextran sulphate in 0.7 M mannitol (pH 5.5) at $25^{\circ}C$ in a shaking water-bath (120 cycles/min, 2.5 cm stroke). The cells liberated by this treatment were washed once with 0.7 M mannitol and incubated for 1 to 2 hrs at $35^{\circ}C$

in 25 ml 2% Onozuka or 5% desalted Meicelase P in 0.7 M mannitol with gentle shaking (50 cycles/min, 2.5 cm stroke). The protoplasts were washed once and were then ready for inoculation with virus.

Inoculation with virus or viral RNA was done either at 25°C or 0°C. Cowpea chlorotic mottle virus (1 µg/ml) and a polycation, e.g. poly-L-ornithine (2 µg/ml) were incubated together for 10 mins in 0.7 M mannitol buffered with 0.01 M sodium citrate (pH 5.2). 10 ml of the mixture was added to a pellet of 2 x 10^6 protoplasts and incubation was continued for a further 10 mins with occasional agitation to prevent the protoplasts sedimenting. The protoplasts were washed three times with 0.7 M mannitol containing 0.01 M $CaCl_2$ and cultured in 10 ml of the medium of Takebe et al. (1968) at 25°C in continuous fluorescent lighting (500 lux). Protoplasts were normally harvested 24-40 hrs after inoculation.

The percentage of protoplasts infected was determined by staining with fluorescent antibody (Motoyoshi et al., 1973a). The amount of virus was measured directly by sucrose-density-gradient analysis of an homogenate of the protoplasts (Motoyoshi et al., 1973b).

The total DNA of Agrobacterium tumefaciens was prepared by the method of Marmur (1961). A lighter satellite fraction of DNA was prepared from this by differential precipitation with M LiCl. Protoplasts were inoculated with bacterial DNA by a procedure like that used for infection with CCMV. The DNA (2 µg/ml) was preincubated at 0°C with poly-L-lysine (2 µg/ml) for 5 mins and then used to inoculate protoplasts in the usual way at 0°C. The washed protoplasts were plated in the medium of Nagata and Takebe (1971) which was solidified with 0.6% agar and contained no hormones. The cultures were incubated at 25°C in continuous light (500 lux). Control cultures contained naphthaleneacetic acid (5 µg/ml) and 6 benzylaminopurine (1 µg/ml).

RESULTS
The conditions for successful inoculation of protoplasts with negatively charged viruses and viral RNA have been known for some years (Takebe and Otsuki, 1969; Aoki and Takebe, 1969; Motoyoshi et al., 1973a). The crucial feature of inoculation is the presence of a polycation, in the absence of which there is little or no infection. The viruses that are positively charged under conditions of inoculation do not require the presence of a polycation for successful infection of protoplasts, for example, pea enation mosaic virus

(Motoyoshi and Hull, 1974) and brome mosaic virus (Motoyoshi et al., 1974a). This is probably because the positively charged virus particle can approach the negatively charged surface of the protoplasts and be absorbed. In the case of the negatively charged viruses and all classes of RNA, which are of course primarily polyanionic, the negative charge must first be reduced by the addition of suitable polycations to permit unimpeded access to the negatively charged protoplast. The polycation that has found most use in the past has been poly-L-ornithine. Table 1 shows however that a variety of polymers of basic amino acids has proved satisfactory with CCMV, and we have used poly-L-lysine in some of the present experiments because it has much lower toxicity than poly-L-ornithine and can be metabolized readily by the protoplasts. There is therefore less difficulty in obtaining good plating efficiencies after protoplasts have been treated with virus or DNA and then used in wall regeneration and cell division studies.

Table 1

The use of polycations to promote infection of
tobacco protoplasts by cowpea chlorotic mottle virus

Protoplasts were inoculated with CCMV (0.5 μg/ml) in the presence of different polycations (1 μg/ml). DEAE dextran is extremely toxic, and an aggregate of CCMV and DEAE dextran was prepared by mixing the two at concentrations of 1 and 2 mg/ml respectively, collecting the precipitate and resuspending at a concentration giving 0.5 μg CCMV/ml. Virus was assayed after 30 hrs culture in the case of DEAE dextran and 40 hrs in the other cases.

Polycation	Molecular weight (x 10^{-4})	% infected protoplasts
Poly-L-ornithine	12	80
Poly-D-lysine	7.0	75
Poly-L-lysine	5.0	56
Poly-L-arginine	5.5	45
DEAE dextran	50	2.8

The percentage infection obtained by these procedures using virus and freshly prepared protoplasts is very high, approaching 100% in many cases. The efficiency of infection with free RNA is always much lower, usually around 10%. The choice of plant material is critical if susceptible protoplasts are required, and each plant species behaves differently. For example, pea is a host for CCMV, but we have not been able to obtain useful levels of infection in protoplasts prepared from plants about 30 days old; much younger plants are almost certainly necessary. Intact cells do not become infected to any worthwhile level by the methods described here, and it is necessary to use a mechanical method of inoculation of the type described by Murakishi et al. (1970, 1971).

Intact virus is relatively stable in the presence of nucleases and inoculation at 25°C is satisfactory. Viral RNA is however rapidly destroyed during inoculation and DNA will also be degraded under these conditions because there are always present damaged protoplasts which release a variety of enzymes into the inoculation medium. Table 2 shows that it is possible to inoculate protoplasts with the RNA of CCMV at 0°C, so that degradation of the nucleic acid can be minimized. Since the process of inoculation is essentially completed within a few seconds of introduction of the infecting agent (Table 3), there is little point when using nucleic acids, and particularly RNA, in persisting with inoculation for more than 5 mins.

Table 2
Inoculation of tobacco protoplasts with
cowpea chlorotic mottle virus at 0°C

Temp. during inoculation (°C)	Time at 0°C after inoculation (hrs)	% infected protoplasts*
25	0	52
0	2	43
0	6	40

*Determined 48 hrs after inoculation.

Polycations should be preincubated with the virus or its RNA for some minutes before inoculation if the maximum efficiency of infection is required (Table 4). This probably allows time for interaction of the particles, reducing net charge, but in the case of viruses with divided genomes it may also permit the formation of small aggregates of polycation and virus. Infection by small aggregates may increase the efficiency of infection considerably. Even at 0°C however, RNA degrades relatively rapidly during preincubation with polycation and

124

Table 3
The effect of the period of inoculation on the
infection of tobacco protoplasts with
cowpea chlorotic mottle virus

Period of inoculation	% infected protoplasts[*]
20 secs	9
2 mins	15
10 mins	18
30 mins	39

[*] Determined 44 hrs after inoculation.

Table 4
The effect of the period of preincubation of
virus and virus RNA with poly-L-ornithine on the
subsequent infection of tobacco protoplasts

1. CCMV

Period of preincubation (mins)	% infected protoplasts[*]
2 mins	2.2
5 mins	27
10 mins	28

2. CCMV RNA

2 mins	7.7
5 mins	7.4
40 mins	0

[*] Determined 44 hrs after inoculation.
Preincubation at $0^{\circ}C$; in all other cases at $25^{\circ}C$.

it is necessary to reduce the period of preincubation to suboptimal levels. A relatively simple procedure has been developed however for both virus and RNA which largely avoids this problem. At relatively high concentrations (1 mg/ml) the negatively charged viruses and nucleic acids form aggregates and start to precipitate when mixed with similar concentrations of the polycations. In the modified procedure, the required amounts of infectious agent and polycation at a concentration of 2 mg/ml are mixed and at once diluted

to the normal concentration of inoculum which is then added to the
pellet of protoplasts. These manipulations are carried out at 0°C
when nucleic acids are used. Table 5 shows that using the RNA of CCMV,
this modification can significantly improve the percentage infection
obtained with the usual method.

Table 5

Comparison of two methods of preincubation of virus RNA and
poly-L-lysine on the subsequent infection
of tobacco protoplasts

Virus RNA (2 µg/ml) and poly-L-lysine (2 µg/ml) were preincubated for 3 mins at 0°C and then used to inoculate
protoplasts at 0°C in the usual way. In the alternative
method RNA and poly-L-lysine were mixed at concentrations
of 1 mg of each/ml, diluted at once to 2 µg/ml and immediately used to inoculate protoplasts. The percentage of
infection was determined 24 hrs after inoculation.

Method of inoculation	% infected protoplasts
Dilute procedure	12
Concentrated procedure	25

ACTIVATION OF THE PROTOPLAST SURFACE.

The treatment the protoplasts receive immediately before inoculation greatly influences the ability of infectious agents to enter the
cytoplasm. It is necessary to centrifuge the protoplasts and resuspend them immediately before inoculation (Otsuki et al., 1972). This
treatment activates the protoplast in some way, possibly at the plasmalemma, to give increased receptivity. The degree of activation required varies widely with the protoplast preparation and in general,
the older the leaf and the plant from which it was taken, the more
activation is required to give satisfactory levels of infection.
Figure 1 shows however that the procedure of Otsuki et al. (1972) can
be repeated to increase the percentage infection. The open blocks
show the effect of centrifuging and resuspending protoplasts in the
original inoculum. The solid blocks show the effect of using successive treatments with fresh inoculum of lower concentration than
that used normally. It is thus possible to increase the penetration
of virus into the protoplasts by very simple techniques. This response is clearly important when attempting to introduce DNA into a
protoplast.

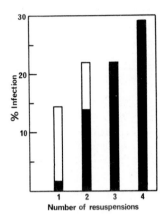

Fig 1 The effect of successive resuspensions on the infection of tobacco protoplasts by cowpea chlorotic mottle virus. In the first experiment (open blocks) protoplasts were resuspended in 20 ml inoculum containing 0.5 μg CCMV and 1 μg poly-L-lysine/ml. They were then spun down and resuspended in the supernatant fluid. In the second experiment (solid blocks) protoplasts were resuspended in successive 10 ml volumes of fresh inoculum containing 0.2 μg CCMV and 0.4 μg poly-L-lysine/ml. The protoplasts were washed and cultured subsequently in the normal way. The percentage infection was determined after 24 hrs culture.

STABILITY OF FOREIGN NUCLEIC ACIDS INSIDE PROTOPLASTS.

Transformation with DNA requires integration of the nucleic acid into the metabolism of the host. There may be some delay in accomplishing this, particularly when using mesophyll protoplasts, which do not synthesize DNA when newly isolated. Little is known however about the stability of unintegrated nucleic acids in the cell. Information about this problem could best be obtained with the aid of a DNA-containing plant virus, for example, cauliflower mosaic virus. Unfortunately it is not at present possible to use this virus in protoplasts and CCMV-RNA has been used instead. Relatively low levels of sodium azide completely inhibit protein synthesis in protoplasts, and the effect may be reversed by washing out the poison (Motoyoshi *et al.*, 1974b). Protoplasts begin to die if poisoned for several hours, but it has proved possible to maintain them in azide for up to 22 hrs and still have about 50% survival. Protoplasts that have been inoculated with CCMV and then cultured in medium containing 0.1 mM sodium azide show no detectable synthesis of virus. When the azide is removed, however, virus multiplication proceeds normally in the surviving protoplasts. Even the longest period of exposure (22 hrs) failed to prevent virus development subsequently (Motoyoshi *et al.*, 1974b). Intact virus can therefore survive for long periods under conditions that do not allow it to be integrated into the metabolism of the protoplast. When however the same procedure was used with viral RNA, the percentage of infected protoplasts fell sharply if the period of exposure to azide after inoculation was greater than about 1 hr. Thus the viral RNA does not appear to survive for very long unless it is integrated into the host's metabolism. It is difficult to extrapolate to DNA, but it is possible that a similar problem may arise when attempting transformation; integration into the host may be too slow in many cases to avoid inactivation of the nucleic acid.

ATTEMPTS TO INDUCE AUXIN INDEPENDENCE IN PROTOPLASTS.

The regeneration of callus from protoplasts of White Burley is completely dependent on the presence of hormones. In general, 6 benzylaminopurine (1 µg/ml) and naphthaleneacetic acid (5 µg/ml) are added to the culture medium to initiate cell division (Watts *et al.*, 1974). In the absence of these there is no division although wall regeneration occurs and the protoplasts survive for 30 or more days. Attempts were made to induce auxin independence by means of DNA isolated from *Agrobacterium tumefaciens*. Two types of experiment were done; in the first, total bacterial DNA was used to inoculate protoplasts in exactly the same way that viral RNA would be used; in the second, a lighter, plasmid fraction of DNA was used. The protoplasts were plated in agar-solidified medium without hormones and cultured in the usual way. There was no evidence of division in either type of experiment, although control cultures with added hormones showed the normal pattern of regeneration with typically 10-20% protoplasts undergoing sustained division.

DISCUSSION

The many attempts to transform plant cells with DNA have produced few positive results. This is perhaps not unexpected since the technical problems are not understood. The studies of infection of callus and protoplasts by viruses indicate that the methods that have been used in the past to introduce DNA into plant cells may have been remarkably inefficient. Cells with intact walls are resistant to infection by tobacco mosaic virus in tissue culture, and Murakishi *et al.* (1970) have shown that the cells must be subjected to mechani-

cal stresses, for example by agitation in a vortex mixer, before
virus can enter efficiently. The same picture has emerged from studies with plant protoplasts; in the absence of trauma produced by
centrifuging and resuspension, levels of infection are rather low.
Further, any delay in introducing the virus after the protoplasts have
been resuspended reduces the percentage of protoplasts that can be infected. It follows that prolonged exposure to solutions of DNA
cannot be expected to be an efficient method of introducing nucleic
acid into cells or protoplasts. The difficulties associated with infection of intact cells also suggests that protoplasts may be a more
suitable starting material for transformation work.

The work with viruses and protoplasts indicates that the following
practical considerations should be taken into consideration during
transformation work with protoplasts.

1. Virtually no nucleic acid will enter the cell in the absence of
suitable concentrations of cation. Motoyoshi *et al.* (1973a) have
shown that there is an optimal balance between the relative amounts
of nucleic acid and polycation; if the amount of nucleic acid is increased beyond a weight ratio of about 10:1 relative to polycation,
the percentage infection falls. There is therefore no advantage in
increasing the input of nucleic acid unless the input of polycation
is correspondingly raised. Polycations are very toxic, and this
means that the maximum levels that can probably be used are of the
order 50 µg nucleic acid and 5 µg polycation/ml. Poly-L-lysine
appears preferable to poly-L-ornithine.

2. The condition of the plant material from which protoplasts are
prepared is critical for inoculation with nucleic acids. The optimal
stages must be determined in preliminary experiments and viruses may
be of use in deciding which are the most susceptible stages. The
amount of nucleic acid absorbed onto the surface of a protoplast is
not a reliable measure of the amount that has entered. Where the
susceptibility is likely to be low, the methods described here will
appreciably improve the amount of nucleic acid entering the protoplasts.

It has been suggested that the induction of tumours by *A. tumefaciens* is mediated by transfer of a fraction of bacterial DNA to the
plant cell. (For a summary, see Heyn, Rörsch and Schilperoort,
1974). The experiments described here indicate that under these conditions fewer than 1 in 10^6 protoplasts was rendered auxin independent by exposure to the DNA of *A. tumefaciens*. It is possible however

that the protoplasts of mesophyll cells are not suitable for this
type of experiment. In the first place they are derived from highly
differentiated cells in which nuclear division has virtually ceased.
There is little synthesis of DNA, and there is no reason *a priori* why
DNA should integrate with the cell genome until the nucleus has been
reactivated and DNA synthesis is possible. This does not occur in
the absence of hormones so that there may be no way of incorporating
foreign DNA unless the protoplasts are given an initial stimulation to
divide. Secondly, even if division is induced by the addition of
exogenous hormones, it will be several days before DNA synthesis begins, and foreign DNA may not survive this period. Protoplasts prepared from rapidly dividing cells may therefore be more satisfactory
material for this type of work.

REFERENCES

Aoki, S., and Takebe, I. (1969). Infection of tobacco mesophyll protoplasts by tobacco mosaic virus ribonucleic acid. *Virology*. 39,
439-448.

Bancroft, J.B., and Flack, I.H. (1972). The behaviour of cowpea chlorotic mottle virus in CsCl. *Journal of General Virology*. 15, 247-251.

Doy, C.H., Gresshoff, P.M., and Rolfe, B.G. (1973). Biological and
molecular evidence for the transgenosis of genes from bacteria to
plant cells. *Proceedings of the National Academy of Sciences, U.S.A.
the United States*

Hess, D. (1972). Transformationen an höheren Organismen. *Naturwissenschaften*. 59, 348-355.

Heyn, R.F., Rörsch, A., and Schilperoort, R.A. (1974). Prospects in
genetic engineering of plants. *Quarterly Reviews of Biophysics*.
7, 35-73.

Johnson, C.B., Grierson, D., and Smith, H. (1973). Expression of
λ p lac5 DNA in cultured cells of a higher plant. *Nature New
Biology*. 244, 105-107.

Ledoux, L. (ed.) (1971). *Informative Molecules in Biological Systems*.
Amsterdam: North-Holland Publishing Co.

Ledoux, L., Huart, R., and Jacobs, M. (1974). DNA-mediated genetic
correction of thiamineless *Arabidopsis thaliana*. *Nature, London*.
249, 17-21.

Marmur, J. (1961). A procedure for the isolation of deoxyribonucleic
acid from micro-organisms. *Journal of Molecular Biology*. 3, 208-218.

Melchers, G. (1972). Haploid higher plants for plant breeding. *Zeitschrift für Pflanzenzüchtung*. 67, 19-32.

Motoyoshi, F., Bancroft, J.B., Watts, J.W., and Burgess, J. (1973a). The infection of tobacco protoplasts with cowpea chlorotic mottle virus and its RNA. *Journal of General Virology*. 20, 177-193.

Motoyoshi, F., Bancroft, J.B., and Watts, J.W. (1973b). A direct estimate of the number of cowpea chlorotic mottle virus particles absorbed by tobacco protoplasts that become infected. *Journal of General Virology*. 21, 159-161.

Motoyoshi, F., Bancroft, J.B., and Watts, J.W. (1974a). The infection of tobacco protoplasts with brome mosaic virus. *Journal of General Virology*. 25, 31-36.

Motoyoshi, F., and Hull, R. (1974). The infection of tobacco protoplasts with pea enation mosaic virus. *Journal of General Virology*. 24, 89-99.

Motoyoshi, F., Watts, J.W., and Bancroft, J.B. (1974b). Factors influencing the infection of tobacco protoplasts by cowpea chlorotic mottle virus. *Journal of General Virology*. (In press).

Murakishi, H.H., Hartmann, J.X., Pelcher, L.E., and Beachy, R.N. (1970). Improved inoculation of cultured plant cells resulting in high virus titer and crystal formation. *Virology*. 41, 365-367.

Murakishi, H.H., Hartmann, J.X., Beachy, R.N., and Pelcher, L.E. (1971). Growth curve and yield of tobacco mosaic virus in tobacco callus cells. *Virology*. 43, 62-68.

Nagata, T., and Takebe, I. (1971). Plating of isolated tobacco mesophyll protoplasts on agar medium. *Planta (Berlin)*. 99, 12-20.

Ohyama, K., Gamborg, O.L., and Miller, R.A. (1972). Uptake of exogenous DNA by plant protoplasts. *Canadian Journal of Botany*. 50, 2077-2080.

Otsuki, Y., Takebe, I., Honda, Y., and Matsui, C. (1972). Ultrastructure of infection of tobacco mesophyll protoplasts by tobacco mosaic virus. *Virology*. 49, 188-194.

Takebe, I., Otsuki, Y., and Aoki, S. (1968). Isolation of tobacco mesophyll cells in intact and active state. *Plant and Cell Physiology*. 9, 115-124.

Takebe, I., and Otsuki, Y. (1969). Infection of tobacco mesophyll protoplasts by tobacco mosaic virus. *Proceedings of the National Academy of Sciences, U.S.A.* 64, 843-848.

Tempé, J. (ed.) (1973). *Colloques Internationaux du C.N.R.S. No. 212. Protoplastes et Fusion de Cellules Somatiques Végétales*. Editions de l'I.N.R.A. Paris.

Watts, J.W., Motoyoshi, F., and King, J.M. (1974). Problems associated with the production of stable protoplasts of cells of tobacco mesophyll. *Annals of Botany.* 38, 667-671.

COWPEA CHLOROTIC MOTTLE AND BROME MOSAIC VIRUSES
IN TOBACCO PROTOPLASTS

J.B. BANCROFT, F. MOTOYOSHI, J.W. WATTS

AND J.R.O. DAWSON

University of Western Ontario, London, Canada.
Institute for Plant Virus Research, Chiba, Japan.
John Innes Institute, Norwich, England.

The conditions required by CCMV and BMV-V5 to infect
protoplasts have been specified. It is fairly clear that
the initial cell-virus association is primarily electrostatic and that subsequent absorption does not have normal
energy requirements and cannot easily be divorced from the
effects of processes which should injure cells. The multiplication of the viruses has been quantified and the production of nucleic acid and protein examined. The relative
rates of production of the four viral RNA's have been described. RNA 3 is made more rapidly than the other RNA's
at 25^0 (although it is not labelled most rapidly) and in
greater quantities than can be encapsidated. RNA 4 is not
made from a complementary negative strand of its own size.
RNA's from two strains of CCMV may multiply in the same
cell. The only translational product so far identified is
the coat protein

Plant virology has entered what promises to be a particularly interesting phase in its development resulting from the availability of
methods, largely developed by Takebe and associates (Takebe and Otsuki, 1969; Aoki and Takebe, 1969) and reviewed by Zaitlin and Beachy
(1974), for obtaining and infecting protoplasts. For the first time,
plant cells can be infected synchronously in a system analogous to
ones used with animal and bacterial viruses. Although current efforts
with plant viruses are still largely centered around the definition of
conditions required for maximum infection, investigations of the events
following infection have also been initiated. We chose to use the
spherical cowpea chlorotic mottle virus (CCMV) and a variant of type
brome mosaic virus (BMV-V5) which multiplies in tobacco. Both viruses
have molecular weights of 4.6×10^6 Daltons and contain single-stranded RNA which is distributed as four species in three nucleoproteins
with slightly different densities. The most dense particles contain

RNA 1 (1.15 x 10^6 D), the least dense RNA 2 (1.0 x 10^6 D) and the particles of intermediate density jointly encapsidated RNA 3 (0.8 x 10^6 D) and RNA 4 (0.3 x 10^6 D). The three largest pieces of RNA are required for infection. The results in this paper are largely concerned with the infection of tobacco protoplasts with these viruses or their RNA's and their subsequent multiplication.

RESULTS

PROTOPLAST PREPARATION AND SOURCE.

The procedures for isolating protoplasts prior to infection with CCMV and BMV-V5 (Motoyoshi *et al.*, 1973; Motoyoshi *et al.*, 1974) are essentially those of Takebe and Otsuki (1969). In short, cells are separated by treatment with a pectinase and the walls are removed with cellulase, the osmoticum being slightly hypertonic (0.7 M mannitol).

The age of the tobacco leaves from which protoplasts are isolated is critical in terms of susceptibility to infection. Variations in the number of infected protoplasts as scored by fluorescent antibody staining (Otsuki and Takebe, 1969) from about 0.2 to 61% occur in preparations from leaves which do not appear to be widely dissimilar (Motoyoshi *et al.*, 1974). The choice of a proper leaf is an essential art which comes with experience. The yields of virus from protoplasts from leaves of different ages does not vary nearly as greatly as does susceptibility.

DOSAGE.

a. *Virus.*

The dosages required to obtain maximum infection efficiencies of tobacco protoplasts with different viruses as measured by fluorescent antibody staining not surprisingly vary somewhat. Maximum efficiency with TMV occurs from 0.1 to 1 µg/ml (Takebe *et al.*, 1971), with cucumber mosaic virus from about 1 to 3 µg/ml (Otsuki and Takebe, 1973) and for CCMV, depending on the experiment, usually from about 0.5 to 2.5 µg/ml (Motoyoshi *et al.*, 1973). These levels are several orders of magnitude less than needed to infect intact tobacco plants efficiently. Under uniform conditions and in terms of particle numbers, TMV would be about 70 times more efficient than CCMV taking into account that its molecular weight is about seven times that of CCMV. However, the spherical virus has a tripartite genome (Bancroft and Flack, 1972), which would reduce the efficiency difference to about a factor of 10. In the case of TMV and CCMV at least, about 1% of the particles added to the protoplasts actually are absorbed (Takebe and Otsuki, 1969;

Motoyoshi *et al.*, 1973). In terms of particle numbers, it has been estimated that 10 to 100 TMV particles are sufficient to infect a protoplast (Hibi and Yora, 1972) whereas about 400 CCMV particles are required (Motoyoshi *et al.*, 1973). In both cases, the measured plating efficiency is only 100 times less than might theoretically be expected which is quite remarkable for a plant virus. The observed value for CCMV, derived from the use of radioactive virus, is an upper limit because it is unlikely that all absorbed particles enter the cytoplasm and function and because many particles may be attached to a poly-L-ornithine molecule, the homopolypeptide being essential for infection with this virus. The procedure for quantifying TMV was based on electron microscopy and probably is a lower limit.

Neither type BMV nor its RNA infects tobacco protoplasts. However, BMV-V5, which is slightly less electropositive than type BMV, infects protoplasts efficiently at 50 to 100 µg/ml. It is about 100 times less efficient than CCMV (Fig 1). In fact, high concentrations of CCMV are less efficient than low ones in causing infection. This probably results from competition for sites on the poly-L-ornithine which itself would be less efficient if saturated with virus particles. Unlike CCMV, neither type BMV nor BMV-V5 precipitates with poly-L-ornithine, which is a reflection of the net positive charge of the viruses under inoculation conditions, nor is poly-L-ornithine required for infection with BMV-V5 (Motoyoshi *et al.*, 1974; Table 1). It would appear from this that partially virus-saturated poly-L-ornithine acts as an attachment vehicle for introducing negatively charged viruses to suitable negatively charged areas on the cell surface which is probably injured by the polycation, but that no such requirement exists for viruses with net positive charges. Such particles adhere in large numbers to the cell surface and a few of them must become attached to just the correct places which are presumably injured and allow infection. The efficiency is lower than with negatively charged viruses because a portal of entry is not created by the agent carrying the virus. The lower the pH at inoculation, the greater the infection efficiency with BMV-V5 (Table 1) but not with CCMV because BMV is more positively charged than at a higher pH. Pea enation mosaic virus also has a net positive charge and does not require poly-L-ornithine (Motoyoshi and Hull, 1974). However, if poly-L-ornithine is present with BMV-V5, the infection efficiency increases (Table 1) and it is simplest to assume that this results from increased cell surface injury providing more infection sites. Clearly, there is no absolute need

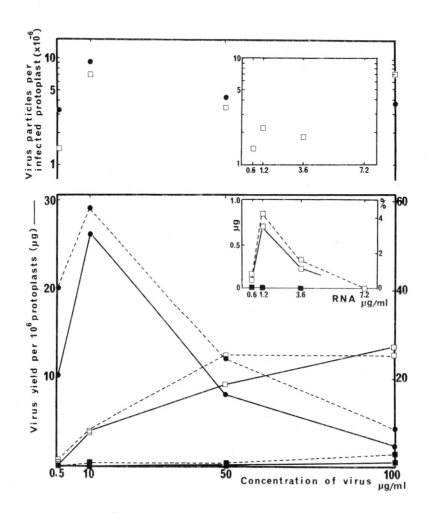

Fig. 1. The response of tobacco protoplasts to different amounts of
wild-type (■) and V-5 (□) brome mosaic virus isolates, their
RNA's (see insets) and cowpea chlorotic mottle virus (●) 44 hrs
after inoculation.

for an agent which induces pinocytosis. All the negative viruses so
far tested require a basic homopolypeptide for infection and it is
reasonable to assume that the initial attachment between the virus
and cell is electrostatic and that only surfaces with a net negative
charge are the sites for infection.

Table 1
The effect of pH and poly-L-ornithine
on the infection of tobacco protoplasts with
brome mosaic virus, V-5.

Inoculum pH	poly-L-ornithine (μg/ml)	Time of infection (hr)	% infected protoplasts [a]
4.7	0	1	0 (0) [b]
4.7	0	24	45 (0)
4.7	1	1	0 (0)
4.7	1	24	77 (68)
5.2	0	1	0 (0)
5.2	0	24	22 (0)
5.2	1	1	0 (0)
5.2	1	24	38 (67)

a) As scored by fluorescent antibody staining.
b) The bracketed percentages are for CCMV (Motoyoshi et al., 1973).

b. RNA.

The use of TMV-RNA as inoculum at concentrations of from 0.5 to 1.5 mg/ml resulted in 3 to 7% infection of tobacco protoplasts (Aoki and Takebe, 1969). CCMV-RNA (Motoyoshi et al., 1973) and BMV-V5-RNA (Fig 1) infect about the same percentage of protoplasts as TMV-RNA at 1.2 μg/ml. Why these RNA's are about 1000 times more efficient than TMV-RNA is not known nor is the reason for the shape of the response curve in Figure 1, which also occurs for CCMV (Motoyoshi et al., 1973), understood. Since the specific infectivities of CCMV-RNA and BMV-V5-RNA are about the same, the difference between the infectivities of the viruses is clearly related to differences in their coat proteins.

INFECTION FACTORS.

a. Inoculation methods and conditions.

Certain variations in when a mixture of CCMV and poly-L-ornithine are added to protoplasts to cause infection are permissible (Motoyoshi et al., 1974). Normally, a two-stage method has been used whereby protoplasts are suspended in 10 ml 0.7 M mannitol and immediately mixed with an equal volume of 0.7 M mannitol buffered with 0.2 M sodium citrate containing 1 μg CCMV and 2 μg poly-L-ornithine/ml. After 10 min at 25°, the cells are collected, washed and cultured. Alternatively, a direct method of inoculation may be used in which the pelleted protoplasts are directly suspended in inoculum constituted as above, diluted with 0.7 M mannitol and treated as before. The latter method may confer a slight advantage for inoculation with virus to fresh protoplasts but doubles the efficiency of infection with RNA.

The direct method has another important advantage in that it allows the use of stored protoplasts in infection experiments. If protoplasts are stored before inoculation by the two-stage method, infection is low, whereas if the direct method is used, infection is comparable to that of non-stored controls (Motoyoshi et al., 1974). The advantage other than logistic of being able to infect stored protoplasts is that the way to superinfection experiments is now open.

It is not permissible to add CCMV and poly-L-ornithine separately to protoplasts if efficient infection is expected (Motoyoshi et al., 1974). The poly-L-ornithine may be neutralized by cell charges if added first and virus adsorbed during pretreatment hinders the basic polymer from affecting the wall at the site of virus adsorption. Nor does injury of cells by whirling or sonication allow infection in the absence of poly-L-ornithine. This is not surprising if negatively cha-rged areas are the infection sites. In addition, it is necessary to preincubate the polymer and virus for at least 5 mins before inoculat-ion. RNA does not show this effect because it is degraded during pro-longed incubation periods even at 0°. The best concentration of poly-L-ornithine is 1 μg/ml.

Negatively-charged viruses will not infect protoplasts in the ab-sence of poly-L-ornithine which is commonly used to aid infection. Protamine sulphate of an unspecified molecular weight has also been re-ported as being effective (Aoki and Takebe, 1969). We have extended the list of useful polycations as shown in Table 2. It is clear that several positively charged polymers are effective and that molecular weights are not critical at least between 50,000 and 120,000. What is critical is that the polymer does not have toxic effects such as those associated with DEAE dextran.

In regard to the kinetics of virus absorption leading to infection, Table 3 shows that infection can occur within 20 sec after addition of virus to the protoplasts. At this time, it is estimated on a basis of radioactivity that x 3 10^{3} particles are attached to each protoplast that becomes infected. If it is assumed that the virus is evenly dis-tributed amongst all protoplasts, then about 300 particles are sufficient to cause an infection. This estimate is in good agreement with that of 400 particles previously reported by Motoyoshi et al. (1973). Because of the low efficiency of absorption, it is difficult to determine the rate constant with any accuracy. However, from 2 mins on, it can be estimated that 1 to 2 virus particles became irreversibly absorbed by each protoplast per second.

Table 2

The effect of various polycations
on the infection of tobacco protoplasts with CCMV.

Polycation (1 µg/ml)	Molecular weight (x 10^4)	% infected protoplasts [a]	Virus particles/ infected protoplast (x 10^6)
Poly-L-ornithine	12	80	8.9
Poly-D-lysine	7.0	75	7.1
Poly-L-lysine	-	56	8.1
Poly-L-arginine	5.5	45	9.2
DEAE dextran	50	2.8	8.2

a) As scored by fluorescent antibody staining 40 hrs after infection except for DEAE dextran which was 30 hrs.

Table 3

Kinetics of CCMV uptake by tobacco protoplasts.

Inoculation time	% infected protoplasts[a]	Virus particles initially absorbed[b] per protoplast	per infected protoplast (x 10^{-3})	Virus particles/ infected protoplast (x 10^{-6})
20 s	9	280	3.0	8.1
2 min	15	340	2.3	4.4
10 min	18	860	4.7	2.8
30 min	39	1,180	3.0	4.1

a) As scored by fluorescent antibody staining 44 hrs after infection.
b) Based on ^{14}C-labelled virus.

There are other factors which affect susceptibility. The pH of inoculation is not as critical with CCMV (Motoyoshi et al., 1973) as it is with BMV-V5. However, infection is high with both viruses at pH 4.7. The optimal pH for infection with RNA is 5.2. Calcium, which is supposed to stabilize cell membranes, reduces susceptibility if present at 10 mM during preinoculation washing of protoplasts (Motoyoshi et al., 1974). Pretreatment with EDTA does not increase susceptibility significantly. It is possible that the different susceptibility of cells from leaves of different ages reflects bound calcium levels.

b. *Infection mechanism.*

There has been considerable discussion as to whether pinocytosis is involved in virus uptake by plant cells. Cocking (1966), Hibi and Yora (1972), Otsuki et al. (1972), and Honda et al. (1974) have suggested pinocytotic uptake whereas Burgess, et al. (1973) have presented evidence to the effect that virus enters cells through damaged membrane areas. Pinocytosis is an energy requiring process which should not be potentiated by cell injury. We have examined virus uptake in terms of these properties rather than on appearance in the electron microscope.

In order to examine the effects of cell injury, protoplasts were repeatedly centrifuged and inoculated. In the first experiment, cells were inoculated by the direct method after one (normal treatment) and two centrifugations and resuspensions in normal inoculum. The treatment increased infection by 35% (Table 4). In the second experiment with protoplasts from the same batch as used in the first experiment, one-fifth the normal concentration of virus and poly-L-ornithine was added successively to protoplasts after successive resuspensions. The number of infected protoplasts increased at each stage to a final level of twice that found in the normal control (Table 4), even though only 80% of the control inoculum was supplied. These results suggest, but do not prove, that infection efficiency is related to mechanical damage.

If a metabolic process such as pinocytosis were involved in virus uptake, the process should be temperature dependent. Yet even prolonged periods at $0°C$ as compared to $25°C$ does not reduce the level of infection markedly (Motoyoshi et al., 1974).

Sodium azide is a potent metabolic inhibitor. Cells treated with the poison at 10^{-4} M lose their ability to incorporate ^{14}C-valine by over 99% within at least 4 hrs (Motoyoshi et al., 1974). The inhibitor may be washed out with the result that the cells regain their normal incorporation rate. The effects of sodium azide on the infection of tobacco protoplasts with CCMV and its subsequent multiplication are listed in Table 5. The first two experiments show that sodium azide did not inhibit virus uptake as measured by the level of protoplast infection. This does not necessarily prove that no energy is required since the size of the ATP pool is not known, but it is indicative of an essentially passive process inconsistent with pinocytosis but consistent with cell injury. The second two experiments show that virus multiplication may be stopped completely or for periods of up to at least 21.5 hrs before multiplication is allowed to resume. Because it is known that virus multiplication starts within at least 7 hrs after inoculation, as will be shown, and because sodium azide should have no effect on uncoating, it appears that the released viral genome can be conserved in dying cells for at least 14.5 hrs.

Table 4

The effect of repeated resuspensions on the infection of
tobacco protoplasts by CCMV.

Expt. No.	Treatment	Inoculum Composition	% protoplasts infected [a]	Virus particles/ infected protoplast $(x\ 10^{-6})$
1	Normal control [b]	0.5 µg virus 1.0 µg PLO	14.4	4.4
	Control centrifuged and resuspended	"	22.0	3.7
2	1 resuspension	0.2 µg virus 0.5 µg PLO	1.6	7.0
	2 resuspensions	"	13.8	3.3
	3 resuspensions	"	22.0	2.8
	4 resuspensions	"	29.0	4.7

a) As scored by fluorescent antibody staining 24 hrs after inoculation.

b) Direct inoculation used throughout.

Table 5

The effect of sodium azide on the behaviour of CCMV
in tobacco protoplasts.

Expt. No.	Duration of azide treatment ($\times 10^{-4}$M)	Period of azide application	% infected [a] protoplasts	virus particles/ infected proto- plast ($\times 10^{-6}$)
1	0 (control)	-	28	6.9
	10 min	during inocu-lation	24	3.8
	30 min	immediately after inocu-lation [b]	24	6.1
2	0 (control)	-	59	2.4
	10 min	preinocula-tion	66	2.1
	10 min	during inocu-lation	59	1.7
	10 min + 10 min	pre + during inoculation	52	2.6
3	0 (control)	-	44	4.7
	45 min	immediately after inocu-lation	46	2.9
	2.25 hr	"	33	1.8
	23.25 hr	"	4.2	-
4	0 (control	-	79	6.1
	6.5 hr	immediately after inocu-lation	59	5.8
	14.5 hr	"	44	3.1
	21.5 hr	"	37	5.4

a) As scored by fluorescent antibody staining at 24 hrs except in experiment 4 which was at 48 hrs.

b) Inoculation required 10 min with CCMV + PLO mixture followed by four washes with azide-free 0.7 M mannitol containing 10^{-4} M $CaCl_2$ as usual; immediately after inoculation actually means about 40 min after the 10 min inoculation period.

Three types of evidence have been presented which are consistent with the idea that virus uptake by plant cells is primarily a physical process. Added to these is the knowledge that plant viruses with net positive charges are incorporated into the cell without poly-L-ornithine, the agent which is supposed to induce pinocytosis. The infection process may be most readily visualized as an accumulation of virus around negatively-charged injured areas on the cell surface, mediated by poly-L-ornithine in the case of anionic viruses, followed by absorption. The membrane may be self-sealing by a process requiring minimal energy. Once the virus is trapped in vesicles, multiplication begins.

MULTIPLICATION.

A number of plant viruses have been shown to multiply in tobacco protoplasts, the most extensively studied being TMV (Takebe and Otsuki, 1969), cucumber mosaic virus (Otsuki and Takebe, 1973), CCMV (Motoyoshi *et al.*, 1973), and BMV-V5 (Motoyoshi *et al.*, 1974). With CCMV, multiplication was originally measured by local lesion assays against a standard of known weight as it was with TMV, but it is really easiest and most accurate to measure CCMV as well as BMV-V5 physically after density-gradient centrifugation. In the most critical growth experiment with CCMV, it was conservatively estimated that 6.4×10^3 times more virus was obtained from a single protoplast 44 hrs after infection than was initially absorbed by it (Motoyoshi *et al.*, 1973). Additional evidence for multiplication of CCMV came from the observations that often about 80% of the cells stained with fluorescent antibodies whereas no fluorescence was observed at zero time, that virus particles were seen to accumulate in infected cells with infection time and that radioisotopes were readily incorporated by the virus. These data leave no doubt that CCMV multiplies in protoplasts. Evidence for multiplication based on increased virus yield for BMV-V5 is not convincing because basic viruses have low plating efficiencies and tend to stick to protoplasts much more so than do negatively-charged viruses because the cells have net negative charges. For example at zero time at pH 4.7 at 50 μg/ml, 2.1×10^6 BMV-V5 particles adhered to tobacco protoplasts that became infected whereas only 7.7 times more virus was recovered 24 hrs later (Motoyoshi *et al.*, 1974). However, it is clear that BMV-V5 multiplied, because the percentage of protoplasts which stained with fluorescent antibodies increased from 0 to 45-70% and the virus RNA incorporated ^{32}P up to 10^7 to 18^8 cpm/mg depending on the experiment. Radioisotope incorporation probably provides the easiest and singly most critical criterion for multiplication in otherwise doubtful cases. None of the above criteria apply to type BMV in protoplasts. The reason for the difference in host range between the two strains of BMV is not known.

As to the kinetics of multiplication, the earliest time that CCMV RNA multiplication has been unequivocally identified by ^{32}P uptake is 7 hrs after infection at $24°$ and the earliest time new mature virus has been detected is 9 hrs after infection, which is about the same as for TMV. BMV seems to behave in about the same way although directly comparable experiments have not been done. The "doubling-time" for these viruses, which is a mechanistically incorrect but nevertheless useful way of considering the rate of virus growth, is in the order of 30 min, as pointed out by Zaitlin and Beachy (1974). After the logarithmic phase which is over by about 24 hrs, virus may be released into the culture medium from dead and dying cells. Released negatively-charged virus cannot infect uninfected cells because no poly-L-ornithine is present in the culture medium but basically-charged viruses presumably can.

The number of CCMV or BMV-V5 particles which accumulates in an infected cell is between 10^6 and 10^7 by 24 to 48 hrs after inoculation. This is an impressive quantity. Virus occupies about 1% of the total cell volume which is about 90% vacuole and has a dry weight of about 10% that of the cell. Yet the cell continues to function.

The viruses are confined to the cytoplasm - they are not associated with the nucleus, mitochondria or plastids of the protoplasts (Motoyoshi *et al.*, 1973). Plant viruses are not generally inhibited by chloramphenicol but are by cycloheximide. With CCMV at least, an abnormal type of distended endoplasmic reticulum is found, the lumen of which contains amorphous electron-dense material (Motoyoshi *et al.*, 1973). This pathological structure may be the seat of virus multiplication and deserves further attention.

Infected protoplasts are routinely incubated in 0.7 M mannitol containing 0.2 mM KH_2PO_4, 1 mM KNO_3, 0.1 mM Mg SO_4, 0.1 mM $CaCl_2$, 1 μ MKI and 0.01 μM $CuSO_4$. No additive such as 2, 4-D, 6-benzyladenine or sucrose has any effect on virus yield (Motoyoshi *et al.*, 1974). The potential of protoplasts for making virus has not yet been improved by manipulating cultural conditions.

It is wise, particularly in labelling experiments, to inhibit the growth of micro-organisms which may contaminate protoplast preparations.

Table 6
The effects of some antibiotics on the
multiplication of CCMV in tobacco protoplasts.

Antibiotic	Concentrated tested	Effect on protoplasts	Multiplication inhibition
CELL MEMBRANES			
Carbenicillin	50-100 µg/ml	none	none
Cephaloridine	300 µg/ml	none	none
Mycostatin	75-250 units/ml	none	none
Rimocidin	10 µg/ml	none	none
Cephaloridine + Rimocidin	300 µg/ml + 10 µg/ml	none	none
Carbenicillin + Mycostatin	100 µg/ml + 25 units/ml	none	none
Amphotercin B + Carbenicillin	2.5-5.0 µg/ml 100 µg/ml	none	75%
30 *S RIBOSOMES*			
Kanamycin	20-100 µg/ml	toxic	complete
Tetracyclin	10-100 µg/ml	none	90%
Gentamicin	5-20 µg/ml	none	none
DNA *DEPENDENT* RNA			
Actinomycin D	40 µg/ml	none	none

Takebe *et al.* (1968) combined the bacteriocide penicillin cephaloridine with the fungicide rimocidin. We routinely use penicillin,
carbenicillin and mycostatin (Watts and King, 1973) because of their
availability and because carbenicillin has a wider antibacterial spectrum than cephaloridine. These antibiotics, which act against membranes, do not affect protoplasts or virus multiplication adversely
either singly or in combination. A list of antibiotics and their
effects on protoplasts and virus multiplication is given in Table 6.
Amphotercin B, which is a polyene that affects cell membranes, in
some way inhibited virus multiplication by 75%, at least in combination with carbenicillin. The antibiotics which affect translation in
procaryotes induced various responses in the protoplasts. Kanamycin
killed the cells, tetracycline inhibited virus multiplication without
reducing the number of cells that became infected and gentamicin had
no effect. Actinomycin D did not inhibit virus multiplication in tobacco, as will be shown in the next section. In spite of the deleter-

ious effects of kanamycin and tetracycline there is a reasonable battery of antibiotics to choose from because the membrane of plant protoplasts differs from those of fungi and bacteria.

VIRAL RNA IN PROTOPLASTS.

a. Single-stranded RNA CCMV-WT.

The synthesis of viral RNA in CCMV and BMV-infected protoplasts has been examined. ^{32}P was usually supplied at about 50 µc/ml at the time of inoculation and the protoplasts were harvested and phenol-extracted from 6 to 44 hrs after inoculation depending on the experiment. The RNA species were separated on 2.6% acrylamide gels after 2.5 to 3 hrs at 4 mA/tube and monitored optically and for counts. The four separate CCMV viral RNA species could easily be identified by radioactivity since the virus multiplied well in the presence of actinomycin D as noted. RNA species 2 and 3 could readily be resolved optically from host components since they migrated between cytoplasmic rRNA's 1 and 2, chloroplast rRNA 1 usually being degraded to give the faster moving components seen in Figure 2. CCMV viral RNA species 1 could be separated from cytoplasmic rRNA 1 so that the former could be quantified optically if electrophoresis was carried on for about 4 hrs, but this was not usually done. BMV species 1 and 2 could be identified optically after a 2.5 to 3 hr run. A host component which coelectrophoresed with viral species 4 was also often found in uninfected protoplasts making optical quantification unreliable.

The results depicting the ratios of CCMV RNA components 1, 2, 3 and 4 based on counts, absorbancies and specific radioactivities are given in Table 7. In experiment 1, the discrepancy between the relative amounts of components 2 and 3 in the treatments without and with actinomycin D is real but is probably not significant in terms of the effect of the antibiotic, since there was considerable unexplained interexperimental variation with this ratio in other experiments in all of which actinomycin D was used. No virus-related counts were observed 6 hrs after inoculation but by 7 hrs (Expt. 4), at which time no virus particles could be detected after density-gradient centrifugation and counting, component 3 was in excess, probably being made most rapidly. This would suggest that an early product is translated from RNA 3. If so, this is probably the 34,500 Dalton protein detected by Shih and Kaesberg (1973) for BMV-RNA 3 in a cell-free system and which may be the virus-specific subunit of the viral replicase (Hariharasubramanian et al., 1973). By 10 hrs after infection (Expt. 3), the ratio of ^{32}P counts among the RNA species reached the level that would be maintained throughout the multiplication cycle, even though at 10 hrs only 10% of the protoplasts that eventually stained at 23 hrs exhibited fluorescence and only about 15% of the RNA was encapsidated. That is, the ratio of RNA species 1:2:3:4 in terms of counts averaged to 0.3: 0.6: 1.0: 0.2 in experiments run at a normal temperature $(24^\circ C)$. The ratio of the optically-identifiable species 1:2:3 averaged 0.2: 0.4: 1.0. If all four RNA species were made at the same rate, then equimolar quantities should be detectable optically. This is clearly not the case. RNA 3 is always present in the greatest quantity at 24° and it must be made most rapidly because of this. But the specific radioactivities indicate that at least RNA 1 and 2 and probably 4 are made more rapidly than RNA 3. This apparent paradox can be reconciled with the data if it is assumed that the pool situation for RNA 3 differs from the others. Alternatively, preferential turnover can be envoked. The former idea is preferable because of the long labelling period and because pool effects should reflect the precursor used. With ^{14}C-adenosine, the CPM ratio of RNA 2: RNA 3 was found to be 1.0: 0.8 whereas that for ^{32}P labelled sister cells it was the opposite. Thus, the most reliable estimates are those obtained from absorbancies.

All of component 3 cannot be encapsidated, since in order to make middle component density virus, RNA components 3 and 4 must be present in equimolar amounts. That they are not is shown most clearly in Figure 3 (corresponding to Expt. 2 (24 hrs), Table 7). At least 50% of component 3 is not encapsidated on the above basis. Further, the molar ratios of RNA's 1 and 2 in purified preparations are about the same, so that it would appear that all of RNA 2 was not encapsidated in this experiment.

b. CCMV-*ts*.

A temperature sensitive (*ts*) mutant of CCMV has been isolated which has a maturation fault and produces free RNA in the inoculated leaves of intact cowpea plants under restrictive conditions (Bancroft *et al.*, 1972). The results found from intact plants were confirmed and extended through the use of tobacco protoplasts. At 24°, both the mutant and wild-type (*wt*) CCMV produced virus and exhibited similar RNA labelling patterns (Expt. 5, Table 7). At 35°, both isolates made RNA but only wild-type produced virus; no *ts* virus nor *ts* coat protein could be detected after examining the contents of ^{14}C-leucine labelled protoplasts. Either the coat protein is not made or is made in small amounts because of translational difficulties at 35° or it is digested

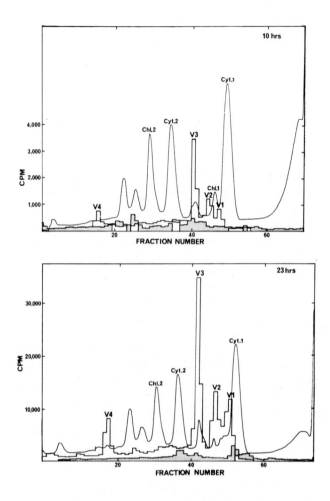

Fig. 2. Polyacrylamide gel electrophoresis patterns of single (left) and double (right) stranded RNA from tobacco protoplasts infected for 10 hrs and 23 hrs with CCMV. ^{32}P was supplied at the time of infection in the presence of actinomycin D. The hatched histograms represent counts from control protoplasts; the open histograms those from virus-infected protoplasts. The labels V1, V2, V3 and V4 refer to the 4 CCMV-RNA species. The curved lines represent optically detectable species, none of which appeared in the *ds* preparations. Cyt. 1 and Cyt. 2 are cytoplasmic ribosomal RNA's; Chl. 1 and Chl. 2 are chloroplast ribosomal RNA's. The unlabelled species are chloroplast RNA 1 degradation products. No optically-detectable viral RNA is found after 10 hrs infection but after

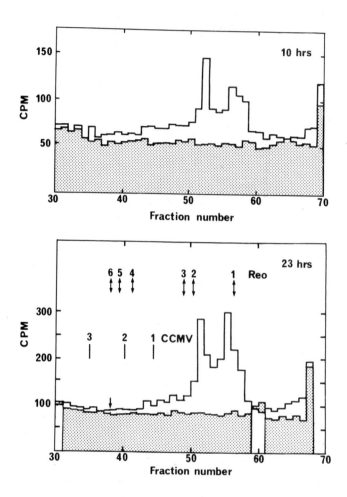

23 hrs, species 2 and 3 are clearly visible under the corresponding histograms. The optical material almost under V3 at 10 hrs
is a chloroplast RNA degradation product which disappears with
protoplast culture time.   The vertical lines labelled Reo on the
figures showing counts obtained after RNAase treatment (10 ng/30 µg
RNA in 2 x SSC for 10 min rt.) give the position of marker *ds*
RNA's from reovirus and those labelled CCMV, the *ss* RNA from CCMV
run at the same time on separate gels. The arrow on the lower
right-hand histogram denotes the position where *ds* RNA 4 would
be found. Electrophoresis for *ss* RNA was for 2.5 hrs and for
ds RNA was for 4 hrs on 2.6% gels, 4 mA/gel.

Table 7

Labelling of the nucleic acids of strains of CCMV
extracted from tobacco protoplasts infected for
various times at various temperatures.

Expt. and strain CCMV	Infection time and temperature	RNA 1 ^{32}P : UV	RNA 2 ^{32}P : UV	RNA 4 ^{32}P : UV
1. WT	24h* : 24°	- : -	0.33 : 0.21	0.32 : -
	24h : 24°	0.20 : -	0.58 : 0.47	0.22 : -
2. WT	6h : 24°	no radioactivity		
	17h : 24°	0.23 : 0.20	0.66 : 0.57	0.17 : -
	24h : 24°	0.28 : 0.23	0.80 : 0.67	0.21 : -
3. WT	10h : 24°	0.21 : -	0.53 : -	0.21 : -
	23h : 24°	0.39 : -	0.50 : 0.22	0.28 : -
4. WT	7h : 24°	0.05 : -	0.17 : -	0.06 : -
	44h : 24°	0.19 : -	0.59 : -	0.18 : -
5. WT	24h : 24°	0.42 : -	0.51 : 0.37	0.19 : -
WT	24h : 35°	0.05 : -	1.72 : 2.70	0.16 : -
ts	24h : 24°	0.49 : -	0.49 : 0.25	0.21 : -
ts	24h : 35°†	0.0 : 0.0	0.66 : 0.45	0.23 : -

BMV		RNA 1 + 2 ^{32}P : UV		RNA 4 ^{32}P : UV
6. BMV-V5	41h : 24°	0.41 : 0.26		0.38 : 0.22
7. BMV-V5	24h : 24°	2.6 : -		0.36 : -

The results are expressed in radioactive counts (^{32}P) and ultra-violet absorption measurements (UV), both relative to RNA 3 = 1.00. Dashes indicate that the measurements are not recorded.

* In all experiments except this one 40 µg/ml of actinomycin D were present.

† No virus is produced by the ts mutant at this temperature but viral RNA is synthesised.

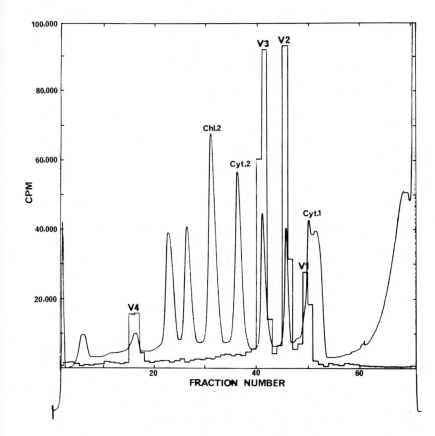

_Fig. 3. Single-stranded CCMV labelled as for Figure 2. Note the large quantity of viral RNA present and that V4 and V3 are clearly not present in optical equimolar amounts. The ratio of V2: V3 is unusually high for incubation at 24°.

in the cell so that it does not migrate as coat protein on polyacrylamide gels. The genetic potential to make coat protein is available since RNA species 3 and 4, both of which carry the coat gene, are clearly present under restrictive conditions, as in RNA 2. Surprisingly, virus RNA 1 could not be detected in $ts/35°$ protoplasts and was found in only very small amounts in $wt/35°$ cells (Expt. 5, Table 7). This may result from changes in pools since $wt/35°$ virus behaves normally in CsCl density gradients. In addition, more RNA 2 than RNA 3 was detected optically in $wt/35°$ protoplasts (Expt. 5, Table 7). This apparently reflects a change in transcription regulation, whereas the counts show a different pattern than found at $25°$, suggesting a tem-

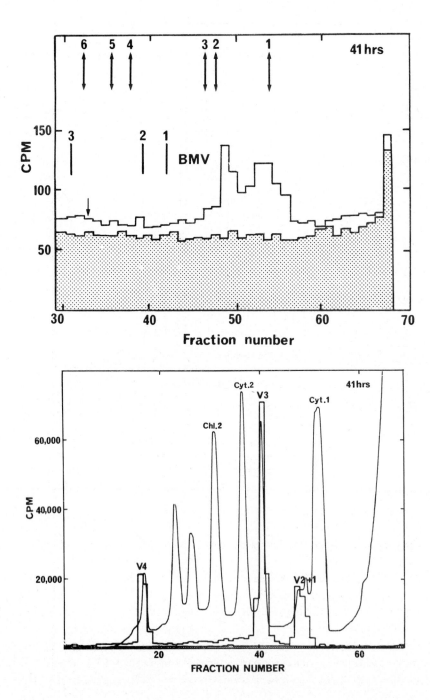

perature effect on the pools.

Since no *ts* CCMV is made at $35°$, the assessment of a phenotypic mixing experiment with protoplasts infected with a mixture of *wt* and *ts* should be relatively easy. It was found that 32 out of 37 lesions tested from mixedly-infected protoplasts contained virus in which *ts* RNA was encapsidated, indicating a surprisingly high level of trans-capsidation. No lesions were obtained from the *ts*-infected $35°$ control.

c. BMV-V5.

The situation with protoplasts infected with BMV-V5 for 41 hrs is shown in Table 7 and Figure 4. Clearly, much more RNA 3 is made on an optical basis than RNA 1 and 2, but as with CCMV, the specific radioactivity of RNA 3 is less than that of the other components. Again as for CCMV, more component 3 is made than is encapsidated. Phillips, Gigot and Hirth (1972), after labelling type BMV-infected barley leaves for 1-6 hrs with ^{32}P, concluded that RNA components 1 and 2 were made more rapidly than component 3 and, in an experiment in which we labelled for 24 hrs, (Table 7, Figure 5) we also found this to be so, contrary to the 41 hrs experiment. We could also detect viral label 8 hrs after inoculation, but the counts were too low for accurate quantification. Neither single nor double-stranded RNA could be detected in protoplasts in which type BMV or its RNA were used as inoculum.

The existence of virus-related low molecular weight (LMC) single-stranded RNA's which are not encapsidated in virus particles has been detected in cells or tissues from plants systemically-infected with TMV (Siegel, Zaitlin and Duda, 1973; MW = 0.35×10^6D), broad bean mottle virus (Romero, 1973; MW = 0.3×10^6D) and BMV (Phillips, Gigot and Hirth, 1972; MW = 0.56×10^6D). The function, if any, of the LMC's is not known but with TMV and broad bean mottle virus, at least, it is clear that they are related to the viral genome. No LMC has been detected in CCMV-infected protoplasts, but a component with a molecular weight of 0.56×10^6D, which is the size of chloroplast ribosomal RNA 2, and with 0.58 times the radioactivity of RNA 3, was found in protoplasts infected with BMV-V5 labelled for up to 24 hrs (Fig 5) but not for 41 hrs (Fig 4).

Fig. 4. Counts of BMV-V5 *ss* and *ds* RNA after 41 hrs infection. Labels are as for Figure 2 except that the hatched histograms represent protoplasts exposed to type BMV.

154

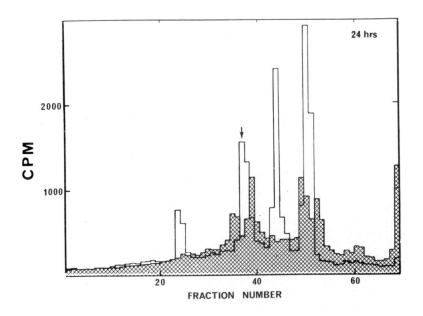

Fig. 5. Counts of BMV-V5 *ss* RNA from tobacco protoplasts infected for
24 hrs. The counts marked with an arrow refer to the anomalous
0.56 x 10⁶D RNA component mentioned in the text.

d. *Double-stranded*

Replicative form (RF) RNA's have been identified after RNAase
treatments for components 1, 2 and 3 (Fig 2, 4). The molecular weights
of the CCMV *ds* strands 1+2, presumably composed of both RNA's 1 and 2,
were 2.25 ± 0.05 x 10^6D and 1.75 ± 0.05 x 10^6D for component 3. Those for
BMV were 2.32 ± 0.12 x 10^6D and 1.75 ± 0.05 x 10^6D, respectively. These
values agree well with expectation. For CCMV, the percentage counts
of *ds*: *ss* after 10 hrs is 5.7% for components 1+2 and 3.7% for 3
whereas they were 1.0% and 0.7% respectively, after 23 hrs. These
figures are consistent with the idea that the *ds* RNA's are formed before the *ss* RNA's but does not specify their function. The approximate
number of *ds* molecules per infected cell at 23 hrs can be estimated
because it is known that each infected protoplast contained 3.5 x 10^6
virus particles. Assuming the three virus particle types found in
CsCl density gradients (Bancroft and Flack, 1972) to be equimolar and
50% encapsidation of components 2 and 3, about 5 x 10^4 *ds* molecules
result as an upper limit on a basis of radioactive counts. But since
the specific radioactivity of the *ds* RNA is undoubtedly at least an

order of magnitude greater than that of *ss* RNA, a reasonable estimate
is that each infected protoplast yields about 5×10^3 *ds* molecules.
This figure is higher than the 100 - 1000 molecules estimated for TMV
in systemically infected plants (Shipp and Haselkorn, 1964), but the
disagreement is not serious, considering the assumptions used in both
cases.

No *ds* RNA was detected that would correspond with what would be expected to occur for component 4. In one experiment on CCMV in which
twice the normal amount of ^{32}P was supplied, the counts of the three
major *ds* components exceeded 1,100 cpm each, yet no trace of *ds* 4
could be seen, nor could it be found for BMV-V5. This result is at
variance with that reported by Lane and Kaesberg (1972), who detected
optical amounts of a putative *ds* form for RNA 4 from barley systemically-infected with BMV, as well as for a minor RNA component. The
significance of our inability to detect a double-stranded equivalent
of RNA 4 centres around how RNA 4 is made. If it were found to exist,
it is likely that component 4 would replicate *via* its own minus strand,
after being derived by nuclease action from component 3, which is
known to duplicate its coat protein synthesizing function. But since
ds 4 has not been detected, it is more probable that it arises directly
from the minus strand of component 3 perhaps by some internally-initi-
ated mechanism, similar in outline to what may occur with vesicular
stomatitis virus (Wild, 1971; Roy and Bishop, 1973). It is less likely that RNA 4 derives directly from RNA 3 plus strands by some degrad-
ative process, because in this case the specific radioactivities would
probably be the same, which is not so. The same considerations are
relevant to the production of the 12S RNA of alfalfa mosaic virus
(Mohier, Pinck and Hirth, 1974).

VIRAL PROTEIN IN PROTOPLASTS.

Very little is known about the proteins synthesized *in vivo* after
infection of plants with viruses. Work with protoplasts, which theoretically should overcome many of the problems encountered with plants,
has so far yielded no unambiguous additional information. Sakai and
Takebe (1972) showed with TMV-infected protoplasts that, in addition
to coat protein, a large protein with a molecular weight of about
140,000 Daltons could be detected in cells infected for 13.5 hrs. Although this size is in the range that might be expected for an undis-
sociated plant viral replicase, it is unlikely to be so under the conditions of separation used. It could represent a protein not coded
for by the virus genome which happens to accumulate to a greater

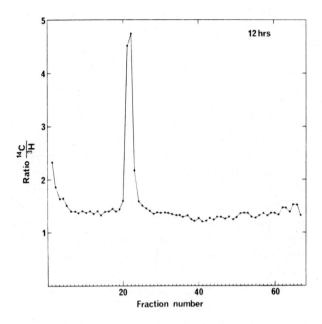

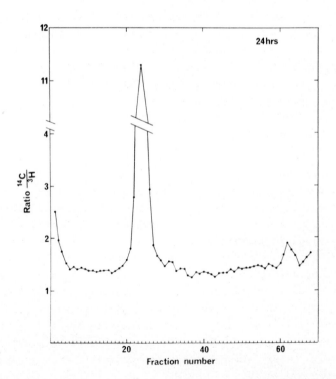

extent in diseased than in healthy cells. Such a response has been discussed by Singer and Condit (1974) who examined the effects of a number of TMV mutants on protein production in tobacco.

Experiments similar to those with TMV have been run with CCMV-infected protoplasts. The single coat protein can be optically identified and difference plots of the counts show only coat protein. More refined comparative procedures in which healthy (^{3}H-leucine) and infected (^{14}C-leucine) cells were mixed, extracted and co-electrophoresed show that, in addition to coat protein, which is clearly evident 12 and 24 hrs after inoculation (Fig 6), a high molecular weight protein can be detected 24 hrs after infection. The late appearance of this protein, along with its size, which is in excess of what could be coded for by a single CCMV RNA species, suggests that it is not involved with virus multiplication. The replicase protein described by Hariharasubramanian *et al.* (1973) for BMV was not detected. Theoretically, about 6 proteins should be made by CCMV but only one has so far been unequivocally detected. Thus an understanding of how the information in the virus genome may function in the successful exploitation of the life processes of plant cells has yet to be achieved. However, plant protoplast-virus systems are still relatively new and continue to promise much.

ACKNOWLEDGEMENT
We wish to thank Ian Flack for excellent technical assistance.

Fig. 6. Plots of the ratios of proteins labelled with ^{14}C-leucine (CCMV-infected) and ^{3}H-leucine (healthy) from tobacco protoplasts 12 and 24 hrs after infection. The large peak at fraction 22 is the coat protein. The small peak at fraction 62 is discussed in the text. Electrophoresis was for 3 hrs on 10% gels, 8 mA/gel.

REFERENCES

Aoki, S., and Takebe, I. (1969). Infection of tobacco mesophyll protoplasts by tobacco mosaic virus ribonucleic acid. *Virology.* 39, 439-
448.

Bancroft, J.B., and Flack, I.H. (1972). The behaviour of cowpea chlorotic mottle virus in CsCl. *Journal of General Virology.* 15, 247-251.

Bancroft, J.B., Rees, M.W., Dawson, J.R.O., McLean, G.D., and Short,
M.N. (1972). Some properties of a temperature-sensitive mutant of
cowpea chlorotic mottle virus. *Journal of General Virology.* 16, 69-
81.

Burgess, J., Motoyoshi, F., and Fleming, E.N. (1973). The mechanism of
infection of plant protoplasts by viruses. *Planta (Berlin).* 112,
323-332.

Cocking, E.C. (1966). An electron microscope study of the initial
stages of infection of isolated tomato fruit protoplasts by TMV.
Planta (Berlin). 68, 206-214.

Hariharasubramanian, V., Hadidi, A., Singer, B., and Fraenkel-Conrat, H.
(1973). Possible identification of a protein in brome mosaic virus
infected barley as a component of viral RNA polymerase. *Virology.*
54, 190-198.

Hibi, T., and Yora, K. (1972). Electron microscopy of tobacco mosaic
virus infection in tobacco mesophyll protoplasts. *Annals of the
Phytopathological Society of Japan.* 38, 350-356.

Honda, Y., Matsui, C., Otsuki, Y., and Takebe, I. (1974). Ultrastructure of tobacco mesophyll protoplasts inoculated with cucumber
mosaic virus. *Phytopathology.* 64, 30-34.

Lane, L.C., and Kaesberg, P. (1971). Multiple genetic components in
bromegrass mosaic virus. *Nature, New Biology.* 232, 40-43.

Mohier, E., Pinck, L., and Hirth, L. (1974). Replication of alfalfa
mosaic virus RNAs. *Virology.* 58, 9-15.

Motoyoshi, F., Bancroft, J.B., Watts, J.W., and Burgess, J. (1973). The
infection of tobacco protoplasts with cowpea chlorotic mottle virus
and its RNA. *Journal of General Virology.* 20, 177-193.

Motoyoshi, F., Bancroft, J.B., and Watts, J.W. (1973). A direct estimate of the number of cowpea chlorotic mottle virus particles
absorbed by tobacco protoplasts that become infected. *Journal of
General Virology.* 21, 159-161.

Motoyoshi, F., Bancroft, J.B., and Watts, J.W. (1974). The infection of
tobacco protoplasts with a varient of brome mosaic virus. *Journal of
General Virology.* (In press).

Motoyoshi, F., and Hull, R. (1974). The infection of protoplasts with pea enation mosaic virus. *Journal of General Virology*. (In press).

Motoyoshi, F., Watts, J.W., and Bancroft, J.B. (1974). Factors influencing the infection of tobacco protoplasts by cowpea chlorotic mottle virus. *Journal of General Virology*. (In press).

Otsuki, Y., and Takebe, I. (1969). Fluorescent antibody staining of tobacco mosaic virus antigen in tobacco mesophyll protoplasts. *Virology*. 38, 497-499.

Otsuki, Y., Takebe, I., Hondo, Y., and Matsui, C. (1972). Ultrastructure of infection of tobacco mesophyll protoplasts by tobacco mosaic virus. *Virology*. 49, 188-194.

Otsuki, Y., and Takebe, I. (1973). Infection of tobacco mesophyll protoplasts by cucumber mosaic virus. *Virology*. 52, 433-438.

Phillips, G., Gigot, C., and Hirth, L. (1972). Rapid labelling of a non-encapsulated RNA of bromegrass mosaic virus. *Federation European Biochemical Society Letters*. 25, 165-169.

Romero, J. (1973). Properties of a slow-sedimenting RNA synthesized by broad bean leaf tissue infected with broad bean mottle virus. *Virology*. 55, 224-230.

Roy, P., and Bishop, D.H.L. (1973). Initiation and direction of RNA transcription by vesicular stomatis virus virion transcriptase. *Journal of Virology*. 11, 487-501.

Sakai, F., and Takebe, I. (1972). A non-coat protein synthesized in tobacco mesophyll protoplasts infected by tobacco mosaic virus. *Molecular and General Genetics*. 118, 93-96.

Shih, D.S., and Kaesberg, P. (1973). Translation of brome mosaic viral ribonucleic acid in a cell-free system derived from wheat embryo. *Proceedings of the National Academy of Sciences, U.S.A.* 70, 1799-1803.

Shipp, W., and Haselkorn, R. (1964). Double-stranded RNA from tobacco leaves infected with TMV. *Proceedings of the National Academy of Sciences, U.S.A.* 52, 401-408.

Siegel, A., Zaitlin, M., and Duda, C.T. (1973). Replication of tobacco mosaic virus. IV. Further characterization of viral related RNAs. *Virology*. 53, 75-83.

Singer, H., and Condit, C. (1974). Protein synthesis in virus-infected plants. III. Effects of tobacco mosaic virus mutants on protein synthesis in *Nicotiana tabacum*. *Virology*. 57, 42-48.

Takebe, I., Otsuki, Y., and Aoki, S. (1968). Isolation of tobacco mesophyll cells in intact and active state. *Plant and Cell Physiology*.

9, 115-124.

Takebe, I., and Otsuki, Y. (1969). Infection of tobacco mesophyll protoplasts by tobacco mosaic virus. *Proceedings of the National Academy
of Sciences, U.S.A.* 64, 843-848.

Takebe, I., Otsuki, Y., and Aoki, S. (1971). Infection of isolated
tobacco mesophyll protoplasts by tobacco mosaic virus. Colloques
internationaux C.N.R.S. No. 193. *Les cultures de tissus de plantes.*
503-511.

Watts, J.W., and King, J.M. (1973). The use of antibiotics in the culture of non-sterile plant protoplasts. *Planta (Berlin).* 113, 271-
277.

Wild, T.F. (1971). Replication of vesicular stomatitis virus: characterization of the virus-induced RNA. *Journal of General Virology.*
13, 295-310.

Zaitlin, M., and Beachy, R.N. (1974). The use of protoplasts and separated cells in plant virus research. *Advances in Virus Research.* (In
press).

SOMATIC HYBRIDISATION

G. MELCHERS, W. KELLER[*] AND
G. LABIB

Max-Planck-Institut für Biologie, Tübingen.

Somatic hybrid plants have been produced by the fusion of
mesophyll protoplasts from chlorophyll deficient, light-
sensitive parents. As a result of complementation they had
a normal chlorophyll content and were light insensitive.
High light intensity was used as a selective agent.

The following conditions have to be fulfilled for a somatic hybridisation:

1. Somatic cells have to be put in a situation such that they are
able to fuse with each other. This condition has been fulfilled for
some time by the production of protoplasts.

2. The frequency of fusions between protoplasts from different
plants has to be high enough to offer a real chance for the production
of hybrid cells. Effective fusion conditions (pH = 10.5, 0.05 M
$CaCl_2$ at $37^\circ C$) have been defined by Keller and Melchers (1973). This
has been confirmed by Potrykus using *Petunia* (unpublished personal
communication), by Binding with *Petunia* (1974), and by Schieder with
Sphaerocarpos (1974). Kao and Michayluk (1974) have shown that polyethyleneglycol (PEG) is a very efficient fusion agent; this has been
confirmed by us (unpublished) and Wallin *et al.* (1974). This second
condition has therefore also been fulfilled.

3. It is necessary that the protoplasts (and the fusion products)
are capable of being cultured to the stage of calli and that the
calli can be regenerated to plants. This condition has been fulfilled
for quite a long time (Takebe *et al.*, 1971; Nagata and Takebe, 1971).

4. Calli which have originated in the course of fusion experiments, or at least the regenerated plants, must have characteristics
that enable one to recognize whether they come from non-fused protoplasts A and B, from the fusion of protoplasts of the parents A+A and
B+B, or whether they are the hybrids A+B[**] sought for.

---

[*] Present address: Canada Department of Agriculture, Research
Branch, Ottawa Research Station, Ottawa K1A 0C6, Ontario.
[**] We have proposed (Melchers, 1974 a,b) that somatic hybrids should
be designated A+B, whereas it is usual to write AxB for sexual hybrids. We have found that the same proposal has been made previously
for the so-called graft hybrid ("burdo") by Winkler (1908).

It would be especially advantageous if (A+B) cells, their calli or the plants regenerated from them have a selective advantage over A, B, A+A and B+B. Any kind of "complementation" in the hybrid between parents having recessive non-allelic genes interfering with growth or development, is suitable for the detection of somatic hybrids.

The first somatic hybrid that was described was the amphidiploid one between *Nicotiana glauca* and *N. Langsdorffii* (Carlson et al., 1972). Sexual hybridisation of these plants is possible and the hybrids spontaneously develop tumours. In callus cultures these tumours are auxin autotrophic. A cursory look at Carlson et al's paper might give the impression that this character was used for the selection of hybrid cells. That is, however, not the case. Out of 2×10^7 mesophyll protoplasts in the single fusion experiment using NaNO$_3$, 33 calli were obtained ("by good fortune" according to Carlson, 1973) on normal Nagata-Takebe-medium (1971) containing auxin and kinetin. On the other hand no fusion could be induced by NaNO$_3$ when non-meristematic cells were used, comparable to those used by Carlson (Power and Cocking, 1971 and our own observations). All the 33 calli turned out to be auxin autotrophic after transfer to an auxin-free medium. But the fact that these calli might appear to have originated from fusions is not a convincing argument, since mutations or especially habituation to auxin autotrophy may have occurred at this frequency. Three regenerated calli produced amphidiploid (*N. glauca* x *N. langsdorffii*) hybrid plants. These three plants could not have originated by mutation or habituation. Neither could they be due to "contamination", in the sense of the usual failures in microbiological methods, since plant cells in culture do not spread like the spores of fungi and bacteria or like viruses through the air. In the Brookhaven laboratory callus cultures of the amphidiploid, sexually produced, hybrids existed. They were used for determining "plating efficiency" with protoplasts out of the sexual hybrids on Nagata-Takebe-medium.

After we had succeeded in regenerating plants from protoplasts, we searched for a system of selection of somatic hybrids after the fusion of somatic cells that would be generally applicable to higher plants. In May 1971 we received from Dr. U. Gerstel, North Carolina State University, Raleigh (we had described our aims when we requested him to send the materials) seeds of *Nicotiana tabacum* "yellow green" and "yellow crittenden". At that time we had no effective fusion method for differentiated cells. We tried to exploit the fusion of protoplasts *in statu nascendi*, which has been observed from time to time. In a big experiment out of which we regenerated 3625 plants, we mixed mesophyll cells of *N. tabacum* "yellow green" and "yellow crittenden" after treatment with pectinase. This mixture was treated with cellulase. In two out of five Erlenmeyer flasks we observed an intensive aggregation of the protoplasts which persisted even after plating on agar. Microscopically it was not possible to ascertain whether or not fusion had taken place. Since the protoplasts were derived from normal plants with 48 chromosomes, the somatic hybrids from fusion should have had 96 chromosomes. We found 35 plants which were definitely "yellow green" probably with 96 chromosomes, one green plant with 48 chromosomes, which had all the characters of "yellow green" apart from green leaves, and with offspring that segregated three "green": one "yellow green". Quite a number of the 180 plants which probably had 96 chromosomes could not be identified definitely as "yellow crittenden". Sexual F_1 hybrids, with their chromosome number raised to 96 by colchicine treatment, did not look very different. While such plants always showed "yellow crittenden" and "yellow green" polyploids in their offspring, the offspring of 27 plants, suspected to be of hybrid character, never segregated for "yellow green" and "yellow crittenden".

It was obviously necessary to find an effective fusion method for differentiated vacuolated cells too. Furthermore, an improvement of the selection system had to be developed. We frequently obtained fusions in a medium of pH=10.5, 0.05 M $CaCl_2$ at $37^{\circ}C$, the osmotic pressure of which was reduced during the preparation of protoplasts by reducing the mannitol concentration from 0.7 M to 0.4 M. After this treatment a major part of the protoplasts divided and produced calli on Nagata-Takebe-medium (Keller and Melchers, 1973).

For the new hybridisation experiments we chose the tobacco varieties "sublethal" (sl_1sl_2) and "virescent" ($vi-A_1$), that we had also obtained from Dr. Gerstel.* We call them "s" and "v". The former are light yellowish-green and their seedlings grow slowly in daylight

---

* We received the first seeds on February 22nd, 1971. At first we had to learn to cultivate these two cultivars. We thank Dr. Gerstel once again for his great help in giving us this useful material.

in a pure inorganic medium and they can hardly grow on soil (Fig 1).
If the plants are cultivated in the shade (i.e. in light of very low
intensity) they grow slowly and bear flowers and fruits after a long
cultivation time. The genotype v is also very light-sensitive in an
inorganic medium, but, after some time, it differs clearly from s
(Fig 1) in that the green parts are more bluish-green. In a little
higher light intensity the youngest leaves become white. We found
that v and s can be cultivated in a low light intensity (800 - 1000
lux) at 28°C and in a relative humidity of 80 - 90%, and that the
isolation of protoplasts from plants grown under these conditions is
frequently successful.

In order to increase the selection pressure in favour of the fusion
hybrids that we hoped for, we produced by "anther culture" (Guha and
Maheswari, 1966; Bourgin and Nitsch, 1967; Melchers and Labib, 1970)
"haploids" of the two varieties v and s. They have 24 chromosomes.

It turned out that the light-sensitivity was considerably reduced
by sugar and amino acids not only in callus culture but also when
seedlings were cultivated on sterile media with sugar only. In an
atmosphere with higher O_2-pressure, on the other hand, the light-
sensitivity - especially of v - was increased greatly. The chlorophyll defect of this variety seems to depend on an intensified photo-
oxidative decomposition of the chlorophyll, that can be partially compensated by organic nutrition. Therefore we reduced the concentration

Fig 1. Seedlings after 1½ months growth in a normal greenhouse in
winter. From left to right: 1. "sublethal" = s 2. (s x v) F_1
3. (v x s) F_1 4. "virescent" = v, demonstrating the light
sensitivity of s and v, the complementation in the hybrids and
the identity of (s x v) and (v x s). (No differences in the
plastom).

of the organic components of the Nagata-Takebe-medium to 1/5 to 1/10, and the concentration of benzyl-adenine to ½ of that present in the auxin-free medium suitable for regeneration of shoots. The calli were cultivated at 8 - 10,000 lux (Osram - L 62 W/32, warm white de luxe) at 28°C. By this means we found fusion hybrids (v + s) between yellowish and light green calli. We found the first one in the beginning of December, 1973 in an experiment started on October 2, 1973 (Fig 2 shows a second one from an experiment on May 22, 1974).[*]

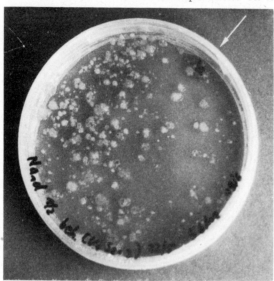

Fig 2. Calluses from a fusion experiment cultivated in Nagata-Takebe-medium with amount of organic constituents reduced to 1/5 , benzyladenine reduced to ½ and without any auxin. v, s, v + v, s + s calli indistinguishable yellowish light greenish. One callus, indicated by an arrow, regenerating dark green and afterwards found to have 48 chromosomes and normal hybrid characters.

The plant has 48 chromosomes. This is a necessary condition for a somatic hybrid originating from "haploid" v and s with 24 chromosomes each. It is, however, not a definite proof, since a spontaneous increase of chromosome numbers in callus cultures is not a very rare phenomenon, and revertants with normal green chlorophyll are possible (see above). But these hybrids are not only normal green and light

[*] Note added in preparing this paper (November 11, 1974): In the meantime 11 fusion hybrids that have arisen independently from each other have been isolated. A detailed paper will be published in *Molecular and General Genetics 135*, 1974.

Fig 3. *Nicotiana tabacum.*
     The first somatic hybrid after fusion of mesophyll· protoplasts
of "haploid" plants of the varieties "sublethal" (s) and "virescent" (v) (cf. Fig 1) (v + s) in the middle between two speci-
mens of the sexual hybrid (v x s), lemonstrating the identity of
the morphological characters.

insensitive like the sexual hybrid of v x s, but they are also similar in all other morphological characters to the sexual hybrids
(Fig 3). Therefore it is most unlikely that they are revertants of
v or s which have had a spontaneous duplication of their chromosome
number in callus culture. The completely convincing proof of the
fact that the first somatic hybrid (v + s) is not a revertant is
furnished by the comparison of its offspring with those of a sexual
hybrid (v x s) (Table 1). For the complementation of chlorophyll deficiency of v and s in the sexual hybrid it does not matter whether
v or s is used as the paternal or the maternal parent in crossing
experiments. We are dealing herewith nuclear genes. Therefore we can
expect a complete identity of the proportions of segregating classes
from the somatic and the sexual hybrids. The situation would be
otherwise if the chlorophyll deficiencies in sexual hybridisation
were transmitted preferentially or only by the mother plant, that is
to say if they were located partially or totally in the plastids.
In *N. tabacum* chlorophyll deficiencies based on plastom mutation are
"albomaculate", i.e. not transferable by the pollen.

     Since chlorophyll deficiency is a character frequently found in
higher plants as a recessive mutation, and since light-sensitive
mutants are not rare, the selection system used here for the first
time can be applied more generally. "Crossfeeding" between v and s
was not observed. Such a phenomenon would restrict the use of complementation between auxotrophic mutants as a selection system for
somatic hybridisation.

Table 1

F_2	green	light green	like v	like s	dead
(v + s)	209	174	101	32	4
(v x s)	108	77	64	20	5

Offspring of the first somatic hybrid (v + s) compared with those of the sexual hybrid (v x s). The existence of the classes "like v" and "like s" also in the offspring of (v + s) proves definitely that the normal green, light-resistant plant cannot be a revertant of v or s.

It was reasonable to study the first somatic hybrids in cases in which sexual hybridisation was possible as well. We know from the work of Kao and Michayluk (1974) and from our own unpublished observations on *Nicotiana tabacum* and *Petunia hybrida* that cell fusion and the subsequent first stages of callus growth are possible even in somatic hybrids, the corresponding sexual hybrids of which are unknown. We do not know yet whether hybrid plants can be regenerated from such hybrid calli. No criteria exist for the prediction of success. We are trying this with the light-sensitive varieties v and s of *Nicotiana tabacum* and with a light green recessive (but unfortunately not really light-sensitive) mutant of *Petunia hybrida*.

REFERENCES

Binding, H. (1974). Fusion experiments with isolated protoplasts of *Petunia hybrida* L. *Zeitschrift für Pflanzenphysiologie*. 72, 422-426.

Bourgin, J.P., and Nitsch, J.P. (1967). Obtention de *Nicotiana* haploids à partir d'étamines cultivés *in vitro*. *Annales de Physiologie Végétale, Paris*. 9, 377-382.

Carlson, P.S. (1973). Towards a parasexual cycle in higher plants. *Colloques Internationaux CNRS*. Nr. 212, Protoplastes et Cellules Somatiques Végétales, Discussion Générale. 548.

Carlson, P.S., Smith, H.H., and Dearing, R. (1972). Parasexual interspecific plant hybridization. *Proceedings of the National Academy of Sciences, U.S.A.* 69, 2292-2294.

Guha, S., and Maheswari, S.C. (1966). Cell division and differentiation of embryos in the pollen grains of *Datura in vitro*. *Nature (London)*. 212, 97-98.

Kao, K.N., and Michayluk, M.R. (1974). A method for high frequency intergeneric fusion of plant protoplasts. *Planta (Berlin)*.

168

115, 355-367.

Keller, W.A., and Melchers, G. (1973). The effect of high pH and calcium on tobacco leaf protoplast fusion. *Zeitschrift für Naturforschung*. 28c, 737-741.

Melchers, G. (1974). Genetik und Pflanzenzüchtung mit mikrobiologischen Methoden. *Planta Medica*. Hippokrates-Verlag Stuttgart, in press (Vortrag vor der Gesellschaft für Arzneipflanzenforschung, Juni 1974). a)

Melchers, G. (1974). Summation: Haploid research in higher plants. *Proceedings of the International Symposium "Haploids in higher plants"* June 1974, Guelph, Canada. b)

Melchers, G., and Labib, G. (1970). Die Bedeutung haploider höherer Pflanzen für Pflanzenphysiologie und Pflanzenzüchtung. Die durch Antherenkultur erzeugten Haploide, ein neuer Durchbruch für die Pflanzenzüchtung. *Berichte der Deutschen Botanischen Gesellschaft*. 83, 129-150.

Nagata, T., and Takebe, I. (1971). Plating of isolated tobacco mesophyll protoplasts on agar medium. *Planta*. 99, 12-20.

Power, J.B., and Cocking, E.C. (1971). Fusion of plant protoplasts. *Science Progress, Oxford*. 59, 181-198.

Schieder, O. (1974). Fusionen zwischen Protoplasten von *Sphaerocarpos donnellii* Aust.-Mutanten. *Biochemie und Physiologie der Pflanzen*. 165, 433-435.

Takebe, I., Labib, G., and Melchers, G. (1971). Regeneration of whole plants from isolated mesophyll protoplasts of tobacco. *Naturwissenschaften*. 58, 318-320.

Wallin, A., Glimelius, K., and Eriksson, T. (1974). Effects of polyethyleneglycol on aggregation and fusion of *Daucus carota* protoplasts. *Proceedings of the 3rd International Congress of Plant Tissue and Cell Culture*, Leicester, July 1974.

Winkler, H. (1908). *Solanum tubingense*, ein echter Pfropfbastard zwischen Tomate und Nachtschatten. *Berichte der Deutschen Botanischen Gesellschaft*. 76a, 595-608.

UPTAKE OF CELL ORGANELLES INTO ISOLATED PROTOPLASTS

I. POTRYKUS

Max-Planck-Institut für Pflanzengenetik,
Projektforschung Haploide,
D-6802 Ladenburg/Heidelberg, Germany.

Protoplasts from higher plants are reviewed as experimental
systems in relation to problems in various fields of plant
science. Genetic modification, the most popular field for
studies with protoplasts at present, may be achieved through
somatic fusion, transfer of nucleic acids, of cell organelles,
or of microorganisms, and through mutation and cellular aggregation. In relation to the transfer of cell organelles, only
some introductory steps have been undertaken. The methods
attempted and the results gained so far by the author in
transplanting nuclei, chloroplasts and tonoplasts are described and discussed.

Protoplasts, as plant cells without cell walls, are a system that
might be useful for many aspects of pure and applied botany. Up to
now protoplasts have been isolated from more than 100 different species
of higher plants and I think it worth noting that approximately half
of these are food plants. However, the isolation stage is only the
introductory step for the use of protoplasts and there are only very
few species which are being studied intensively. Let us look at what
protoplasts have been used for and for which problems they are thought
to be a very helpful system.

First of all protoplasts can be induced to produce fertile plants.
This has been successfully achieved with five different genera -
Nicotiana (Takebe *et al.*, 1971), Daucus (Grambow *et al.*, 1972), Petunia (Durand *et al.*, 1973), Asparagus (Bui-Dang-Ha and Mackenzie, 1973)
and Brassica (Karta *et al.*, 1974). Since it is possible to isolate
10^6 to 10^7 protoplasts from 1 gm of leaf tissue and to induce - with
an optimal method - more than 50% of them to form growing plants, this
should be an ideal method for cloning and for the propagation of rare
genotypes (Fig 1).

In the field of cell biology some publications have appeared on
membrane properties, on the action of chelating agents and of deter-
gents, antibiotics, herbicides, and phytohormones. Papers on ultrastructure have been numerous (Cocking, 1972). However, little if any

170

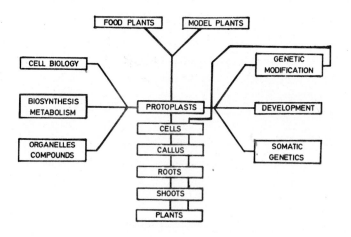

Fig. 1. Use of protoplasts in plant science.

studies have been carried out in interesting fields such as 1) the localization of synthetic loci, of hormone receptors, and of phytochrome, 2) symbiosis, 3) relationship between pathogenic organisms and plant cells, and 4) tumour protoplasts.

In experiments on problems of biosynthesis or metabolism, protoplasts might have a lot of advantages over studies with complete plants or organs:

1) It would be possible to use populations of homogeneous differentiated cells e.g. from the palisade layer of a leaf.
2) The experimental treatment would reach all cells at the same time with the same intensity.
3) The agent might act with a defined intensity or concentration.
4) It would reach the cells unchanged and at a defined time.
5) The entrance into the cell might be easier because of the missing wall.
6) It would be possible to measure the reaction of the cell directly.
7) The reaction products might be isolated more quickly, with less

intefering compounds, and at a higher purity.
It might be of special interest to study the metabolic and biosynthetic potentials of cells in various states of differentiation. Up to
now there are some 20 publications in which the first steps have been
undertaken to make use of the advantages offered by protoplasts (Hoffman *et al.*, 1974).

Similar advantages might apply to protoplasts in the area of cell
organelle isolation (Hoffman, 1973) and the isolation of unstable compounds such as mRNA and tRNA. As protoplasts simply "release" their
contents in a hypotonic environment, the breakdown of the cells can
be carried out far more gently and more quickly than is possible with
the relatively rough methods used for extractions from whole plants.
The consequence might be that it should be possible to get more pure
and less disturbed fractions, and this might be especially important
in relation to very sensitive compounds. In this field only 6 publications exist.

Analogous to the exciting experiments with animal cells, it should
be possible to carry out cell genetics starting with protoplasts: for
example studying gene complementation, compatibility, genetic tumours,
trying to identify linkage groups and genes, analysing gene activation
and inactivation, studying the genetics of cell organelles and nucleo-
cytoplasmic interactions, - these should be all possible.

In the field of development it is especially attractive to attack
problems such as 1) inhibition of cell wall synthesis, 2) inhibition
of pectin synthesis, 3) cell synchronization, 4) embryogenesis, 5)
karyokinesis and cytokinesis, 6) differentiation and dedifferentiation,
and to study the developmental potentials of cells in various states
of differentiation. These problems are closely connected with the
field we are dealing with in this symposium: the genetic modification
of the plant cell. This is true in two respects: to solve the first
four problems would mean direct help for experiments on the genetic
modification of the plant cell. To solve the others would be of great advantage for the regeneration of plants from different genera.

Genetic modification of the plant cell is the field on which most
research efforts are now being focused (Cocking, 1972) (nearly 60 publications up to now). This goal may be approached in different ways
(Fig 2): 1) Somatic fusion (Keller and Melchers, 1973), 2) transfer
of nucleic acids (directly or by the use of bacterial or viral agents)
(Ohyama *et al.*, 1972), 3) transfer of cell organelles (Potrykus and

172

Hoffman, 1973) (nuclei, plastids, mitochondria, ribosomes, tonoplasts, membranes, cytoplasm), 4) transfer of microorganisms (Takebe and Otsuki, 1969) (blue green algae, fungi, bacteria, viruses), 5) mutation, and possibly 6) aggregation of cells from different species and genera (chimeras might be obtained from such cultures).

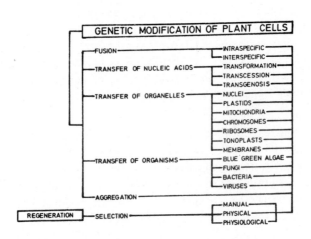

Fig. 2. Possibilities for genetic modification with protoplasts.

As the desired events will in most cases be rare, the development of appropriate selective systems (physiological, physical, manual) are important as well. The most talked about success in fusion has been gained so far by Carlson (Carlson et al., 1972), but his methods have been subjected to criticism. Recently Melchers (1974) has succeeded in obtaining a fertile intraspecific fusion hybrid which has been confirmed by seed progeny tests. Following this report Professor Gamborg will report on his experiments with protoplasts from different genera and even from different families (Gamborg et al., 1974). Another successful field is the transfer of viruses, where the replication and expression of the virus genome in the plant protoplasts has been reported by Doctor Takebe (1974). Besides some experiments on fusion

(Potrykus, 1973a) and on transformation (together with Hess and Hoffmann) I have done some initial experiments on the transfer of cell organelles (Potrykus, 1973b; Potrykus and Hoffman, 1973) on which I
shall now report.

ISOLATION OF CELL ORGANELLES FROM PROTOPLASTS.

Of the existing possibilities for isolating organelles from protoplasts, only a few have been successfully tried: the isolation of
interphase nuclei (Hoffman, 1973), chloroplasts (Potrykus, 1973b) and
tonoplasts.

a. Isolation of nuclei.

Nuclei were isolated from mesophyll protoplasts of Petunia and
Nicotiana tabacum and stained with ethidium bromide according to Hoffmann (1973) by resuspending a sample of protoplasts in equiosmolal
$CaCl_2$, mixing them with an equal volume of "nucleus isolation medium"
(4% Triton X-100 in 0.20 M NET sucrose, pH 8.0, where NET is 0.1 M
NaCl, 0.01 M EDTA, 0.01 M Tris) and stirring them for about 10 mins
at $4^{\circ}C$. The nuclei were gathered by pelleting them down at 1000 x g
for 10 mins at $4^{\circ}C$. The nuclei were then washed in 0.25 M NET sucrose
and were run through a gradient of 28% sucrose over 60% sucrose (both
in NET) at 25000 x g, $4^{\circ}C$ for 45 mins. The nuclei were then stained
with 0.05% ethidium bromide in 0.16 M $CaCl_2$, pH 5.8, $20^{\circ}C$, for 5 mins.

b. Isolation of chloroplasts.

Wild type chloroplasts have been isolated from mesophyll protoplasts
of different species by stirring the protoplasts in mannitol of decreasing osmolarity. The chloroplasts were collected at about 1000 x g
and resuspended in 0.35 M mannitol or 0.21 M sodium nitrate, which is
equiosmolal to 0.35 M mannitol.

c. Isolation of tonoplasts.

Tonoplasts have been isolated from protoplasts from petals of
Nemesia strumosa, Torenia fournieri and *Petunia hybrida* by treating
the protoplasts in isoosmolal mannitol with Driselase, an unpurified
cellulase (Kyowa Hakko Kogyo Company, Tokyo) which contains some pro-
teases. Driselase is obviously far more aggressive against the plasmalemma than against the tonoplast.

TRANSPLANTATION OF CELL ORGANELLES INTO PROTOPLASTS.

The situation concerning the isolation of cell organelles from protoplasts is true also for the transplantation of cell organelles into
protoplasts: the existing possibilities have hardly been fully exploi-
ted.

a. Transplanation of nuclei.

For the transplantation of nuclei, the membrane modifying action of lysozyme was combined with the fusion inducing action of centrifugation and weak deplasmolysis: freshly isolated protoplasts (from Petunia, Nicotiana, *Zea mays*) were suspended in an isotonic solution of mannitol to a titre of 10^6 per ml. Stained nuclei were suspended (10^6 per ml) and stored for at least 1 hr in a solution of 0.03% lysozyme (Serva 28260) pH 5.8, in mannitol which was slightly hypotonic to the protoplasts. At first 1 ml of the protoplast suspension was sedimented in a narrow swing out centrifuge tube (r_{max} 12.5 cm) for 5 mins at 140 x g (Fig 3).

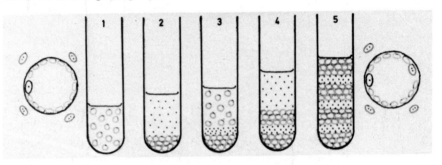

Fig. 3. Transplantation of nuclei into protoplasts by layer-centrifugation, lysozyme and deplasmolysis.

1. Protoplasts 10^6/ml in isotonic mannitol, centrifuge 5 mins at 140 x g. 2. Discard supernatant, layer carefully with nuclei 10^6/ml in hypotonic mannitol + 0.03% lysozyme, centrifuge 5 mins at 140 x g. 3+4. Carefully remove supernatant and repeat 1+2 until pellet consists of several alternating layers of protoplasts and nuclei. 5. Discard supernatant, layer carefully with hypotonic mannitol, centrifuge 30 mins at 140 x g, store at 4°C for 2 hrs.

The supernatant was carefully discarded and 1 ml of the nuclear suspension was carefully layered over the pellet and was centrifuged onto the top of the pellet at 140 x g for 5 mins. Again the supernatant was carefully removed and the pellet was layered by another 1 ml of the protoplast suspension. This was continued until the pellet consisted of 4 layers of protoplasts alternating with 3 layers of nuclei. Then the centrifuge tube was filled with hypotonic mannitol, centrifuged at 140 x g for 30 mins and stored for 2 hrs at 4°C. Thereafter the pellet was very carefully resuspended in isotonic mannitol for observation under the fluorescence microscope. It could be seen that most of the stained nuclei were floating freely and that some of them adhered to the plasmalemma of the protoplasts. However, approximately

0.5% of the protoplasts contained from 1 to 5 stained nuclei together with their own nucleus. This could be confirmed by rolling the protoplasts under the cover slide.

At present we know nothing of the fate of nuclei in the host cell. The experiments have not yet been carried out under aseptic conditions and we have not yet performed experiments using selection methods to obtain a knowledge of the biology of the system comprising engulfed nucleus + host cell.

b. Transplantation of chloroplasts.

In addition to experiments with nuclei, chloroplasts could be transplanted by a similar technique to that already described. However, better results could be obtained with the following method (Fig 4).

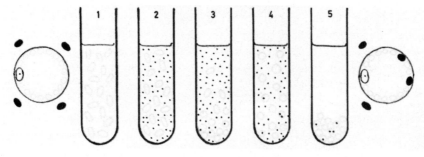

Fig. 4. Transplantation of chloroplasts into protoplasts by co-incubation of protoplasts in "*statu nascendi*" with chloroplasts at low speed centrifugation.

1. Cells after 60 mins in hypotonic mannitol, 2% cellulase pH 5.4, 20°C. 2. Mixed population of single cells and of chloroplasts in isotonic sodium nitrate, 2% cellulase pH 5.4, 20°C, 5 mins at 10 x g. 3. Protoplast in "*statu nascendi*" + chloroplasts, resuspend and centrifuge again 5 mins at 10 x g. 4. Repeat every 5 mins until all single cells have been converted into protoplasts. 5. Wash free of chloroplasts and observe under light microscope.

Pure populations of albino single cells were isolated from an extra-chromosomally inherited variegating *Petunia hybrida* which develops pure white shoots and leaves. Cells were obtained by peeling off the lower epidermis of leaves and incubating the tissue for 15 mins in 2% pectin-ase (Serva 31660) in 0.35 M mannitol, pH 5.4 at 30°C, and shaking the tissue with glass beads. These single cells can be converted into protoplasts if incubated in 2% cellulase (Onozuka SS) in 0.35 M manni-tol or in 0.21 M sodium nitrate for 45 mins at 30°C or for 75 to 90 mins at 20°C. Protoplasts in "*statu nascendi*" accumulate after 60 to

90 mins following transfer of the single cells into the cellulase.
Single cells at a stage where the first protoplasts appeared, were
transferred into a suspension of chloroplasts in 2% cellulase in
0.21 M sodium nitrate, pH 5.4. The mixed suspension (about 10^4 x ml^{-1}
cells and 10^5 x ml^{-1} chloroplasts) was incubated in a swing out rotor
at low speed (about 10 x g) at 20°C for 15 to 30 mins. Every 5 mins
the pellet was resuspended. When nearly all cells had been converted
into protoplasts, the mixed population was washed free of chloroplasts
and the protoplasts were examined in a light field microscope for protoplasts containing white and green plastids. Mixed protoplasts could
be found at a rate up to 0.5%, containing 1 to 20 chloroplasts apart
from the normal 50 to 100 white plastids. To be sure that they represent an uptake of ch loroplasts, two possible sources of error have to
be excluded: first, the possibility of the existence of mixed cells
in the tissue from which the protoplasts have been isolated. This has
been done by intensive microscopic observation and through the experience of years of intensive genetic studies with this mutant. Second,
the chloroplasts might merely adhere to the membrane of the protoplasts
This has been checked in a manner analogous to that used for the uptake of nuclei, by rolling the protoplasts under the cover slide.

c. *Transplantation of tonoplasts.*

Transplantation of tonoplasts could easily be achieved by using
fusion methods: sodium nitrate, or Ca^{2+} + high pH + high temperature
according to Keller and Melchers (1973). The coloured tonoplasts isolated from petal protoplasts can be recognized easily within the host
protoplast. The same is true for white tonoplasts isolated from white
petals and having been taken up into coloured protoplasts from coloured petals. Since more than 5 tonoplasts of a size of more than one-
third of the diameter of the host protoplast had been taken up sometimes, some problems arise concerning the mechanism of uptake. Up to
now it has been generally supposed that cell organelles are engulfed
by an infolding of the plasmalemma, so that they are enclosed by the
host cell envelope. But the surface of the membrane of a protoplast
will surely not be large enough to cover 5 or even more tonoplasts of
such a large size entering the protoplast?

OTHER REPORTS ON TRANSPLANTATION OF CELL ORGANELLES.

To my knowledge there is only one other report on the transfer of
cell organelles into plant protoplasts. Carlson (1973a, b) reported
very briefly and generally, without giving any details, that he had
cultured albino protoplasts isolated from an extrachromosomally inher-

ited, variegating albino mutant from *Nicotiana tabacum* together with
wild type chloroplasts, and that he regenerated green plants from
this experiment. No albino or variegated plant were reported. From
this result he concluded, that wild type chloroplasts had been taken
up into the protoplasts, that they had propagated there, and that the
cells had given rise to green plants. Though nearly two years have
passed since the first report on this exciting experiment appeared,
as far as I know, no details concerning the experiment appear to have
been published yet. I would like to discuss two general doubts concerning the conclusions drawn by Carlson.

1. From all experiences with extrachromosomally inherited mutated
white plastids in higher plants, one has to expect white, variegated
and (very seldom) green plants. If a population of albino protoplasts
is treated with chloroplasts, the chloroplasts are taken up into the
protoplasts and propagate there. Even if one thinks of a selective
growth advantage for cells containing chloroplasts - an assumption
for which no indication exists - one has also to expect at least some
albino and variegated types. Unfortunately Carlson did not report on
a control experiment in which he should have cultured the so-called
albino protoplasts without additional chloroplasts. That is a pity,
because there might be a simple explanation for the fact that he got
green plants by culturing white protoplasts.
2. Then if chloroplasts were not taken up, Carlson's results have to
be explained in another way. White leaves or shoots of extrachromosomally inherited variegating plants are in many cases periclinal chimeras of the type "green/white/white". In that case the plastids of the
epidermal cells are only phenotypically "white" because of their position, but they are genetically "green". In the case of regeneration,
such epidermal protoplasts will naturally grow up to a green plant
without any uptake of chloroplasts. If a mixture of protoplasts from
such a leaf was used, one has to expect green and white plants. Even
variegated ones may not be a proof of the uptake of chloroplasts, for
sectorial or periclinal chimeras can arise from a mixed callus which
might be formed, if genetically white and phenotypically white protoplasts form a joint callus, and plantlets arise by a combination of
more than one cell. Variegated plants must be shown to contain mixed
cells, containing green and white plastids, to exclude further doubt.
But even then, the variegated plant might have been regenerated from
a fusion product of a genetically white and a genetically green protoplast. In such a field it is very important to follow the fate of

transplanted chloroplasts in individual protoplasts cultured in micro-chambers.

I agree completely that it will be a great advance to handle plant cells like microorganisms - but if we do so, we have to interpret the results carefully, and also ensure that proper control experiments are carried out.

ACKNOWLEDGEMENTS

I gratefully acknowledge the facilities and the advice given by Professor Hess, Institut für Botanische Entwicklungsphysiologie, Universität Hohenheim, D-7000 Stuttgart 70, Emil-Wolffstr. 20, where this work was carried out. The contributions of F. Hoffmann are also appreciated.

REFERENCES

Bui-Dang-Ha, D., and Mackenzie, I.A. (1973). The division of proto-plasts from *Asparagus officinalis* and their growth and different-iation. *Protoplasma*. 78, 215-221.

Carlson, P.S. (1973a). Towards a parasexual cycle in higher plants. *Colloques Internationaux de Centre National de la Recherche Scientifique*. 212, 497-505.

Carlson, P.S. (1973b). The use of protoplasts for genetic research. *Proceedings of the National Academy of Sciences, U.S.A.* 70, 598-602.

Carlson, P.S., Smith, H.H., and Dearing, R.D. (1972). Parasexual inter-specific plant hybridization. *Proceedings of the National Academy of Sciences, U.S.A.* 69, 2292-2294.

Cocking, E.C. (1972). Plant cell protoplasts - Isolation and develop-ment (1972). *Annual Review of Plant Physiology*. 23, 29-50.

Durand, J., Potrykus, I., and Donn, G. (1973). Plantes issues de proto-plastes de Petunia. *Zeitschrift für Pflanzenphysiologie*. 69, 26-34.

Grambow, H.J., Kao, K.N., Miller, R.A., and Gamborg, O.L. (1972). Cell division and plant development from protoplasts of carrot cell suspension cultures. *Planta (Berlin)*. 103, 348-355.

Gamborg, O.L., Constabel, F., Kao, K.N., and Ohyama, K. (1974). Plant protoplasts in genetic modification and production of intergeneric hybrids. This Symposium.

Hoffmann, F. (1973). Die Aufnahme doppelt-markierter DNA in isolierte Protoplasten von *Petunia hybrida*. *Zeitschrift für Pflanzenphysiol-ogie*. 69, 249-261.

Hoffmann, F., Kull, U., and Potrykus, I. (1974). Manometrische Atmungsmessungen an Protoplasten von Petunia. *Biologie Plantarum*, in press.

Karta, K.K., Michayluk, K.N., Kao, K.N., Gamborg, O.L., and Constabel, F. (1974). Callus formation and plant regeneration from mesophyll protoplasts from rape plants *(Brassica napus)*. *Plant Science Letters*, in press.

Keller, W.A., and Melchers, G. (1973). The effect of high pH and calcium on tobacco leaf protoplast fusion. *Zeitschrift für Naturforschung*. 28c, 737-741.

Melchers, G. (1974). Summary remarks on haploids in higher plants. *Proceedings of the International Symposium on Haploids in Higher Plants, Guelph, Canada*.

Ohyama, K., Gamborg, O.L., and Miller, R.A. (1972). Uptake of exogenous DNA by plant protoplasts. *Canadian Journal of Botany*. 50, 2077-2080.

Potrykus, I. (1973a). Isolation, fusion and culture of protoplasts from Petunia. In *Yeast, Mould and Plant Protoplasts*. Academic Press, London - New York, 319-330.

Potrykus, I. (1973b). Transplanation of chloroplasts into protoplasts. *Zeitschrift für Pflanzenphysiologie*. 70, 364-366.

Potrykus, I., and Hoffmann, F. (1973). Transplanation of nuclei into protoplasts of higher plants. *Zeitschrift für Pflanzenphysiologie*. 69, 287-289.

Takebe, I. (1974). Replication and expression of tobacco mosaic virus genome in isolated tobacco leaf protoplasts. This Symposium.

Takebe, I., Labib, G., and Melchers, G. (1971). Regeneration of whole plants from isolated mesophyll protoplasts of tobacco. *Naturwissenschaften*. 58, 318-320.

Takebe, I., and Otsuki, Y. (1969). Infection of tobacco mesophyll protoplasts by tobacco mosaic virus. *Proceedings of the National Academy of Sciences, U.S.A.* 64, 843-848.

PLANT PROTOPLASTS IN GENETIC MODIFICATIONS AND PRODUCTION OF INTERGENERIC HYBRIDS

OLUF L. GAMBORG, F. CONSTABEL, K.N. KAO AND
K. OHYAMA

*Prairie Regional Laboratory,
National Research Council,
Saskatoon, Sask. Canada S7N OW9.
NRCC No. 14385*

Protoplasts can be isolated from many plant species by treatment with commercial enzyme preparations. Cell regeneration and induced division is possible with protoplasts from plants of different genera. Intergeneric fusion can be induced by treatment with high concentrations of polyethylene glycol (PEG). The protoplasts agglutinate and fusion occurs when the concentration of the PEG is slowly reduced. Heterokaryocyte production reaches 30%. Soybean-barley, pea-carrot, soybean-rapeseed and other heterokaryocytes divide. Chromosomes of both parental species can be observed in the hybrid cells. Transformation experiments with protoplasts have shown that DNA is taken up as the double-stranded form. Protoplasts from a few plant species can be cultured and the cells induced to differentiate and form complete plants. Because of the recent improvements in protoplast technology and the development of a suitable cell fusion procedure, more rapid progress may be expected in the pursuit of developing intergeneric hybrid plants.

Plant protoplasts provide experimental material for many fundamental and applied aspects of plant science. A rapidly increasing number of reports have appeared describing the use of protoplasts in studies on cell walls and membranes, and on a variety of physiological and biochemical investigations (Tempè, 1973; Fowke et al., 1973, 1974a, b; Kanai and Edwards, 1973). Many of the earliest reports on protoplasts dealt with uptake of virus particles (Cocking, 1970; Takebe and Otsuki, 1969). More recently, significant progress has been made in developing techniques for employing protoplasts in genetic transformation, and in somatic hybridization by fusion (Carlson et al., 1972; Chaleff and Carlson, 1974; Ohyama et al., 1972; Gamborg and Miller, 1974; Keller and Melchers, 1973; Kao and Michayluk, 1974; Hoffmann and Hess, 1973; Holl et al., 1974).

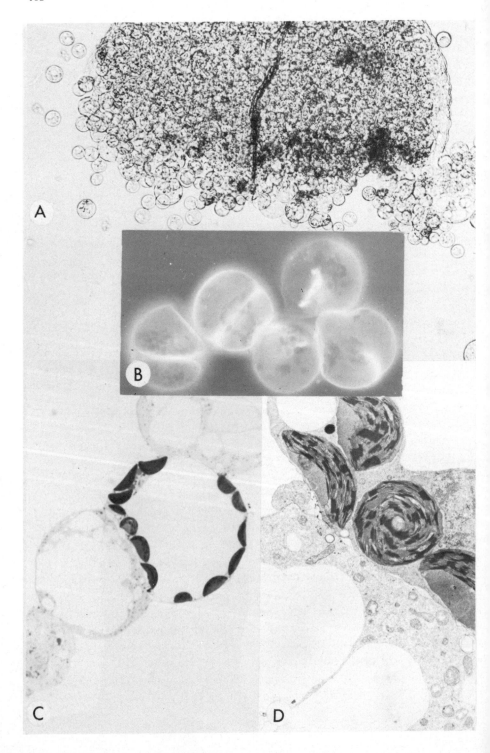

PROTOPLAST ISOLATION AND CULTURE.

Protoplasts have been obtained from numerous plant species, from cultured cells, leaves and other plant organs (Nickell and Heinz, 1973). Commercial preparations of cellulases, hemicellulases and pectinases effectively remove the cell walls and release the protoplasts within a few hours (Constabel, 1974; Nagata and Takebe, 1971; Driselase, 1974; Gamborg et al., 1973). Although the protoplasts can be cultured for many days they are often difficult to induce to divide. The successful survival and subsequent growth and division may depend on the physiological state of the plant material, the isolation conditions as well as the culture medium and environment (Constabel et al., 1973; Pelcher et al., 1974; Kartha et al., 1974a; Fowke et al., 1973; Fowke et al., 1974b). Rapidly growing cell cultures maintained by subculture three times weekly provide a very suitable source of protoplasts. The yields of protoplasts from leaves may depend upon light and temperature conditions during plant growth. The type of enzyme and the sequence of treatment also influence the success of isolation (Constabel, 1974; Fowke et al., 1974b; Driselase, 1974). During culture the prevailing osmolality (0.3-0.5), pH (5.5-6.2), carbon source (glucose) and kind and amount of growth regulators are critical (Kao et al., 1973; Constabel, 1974). The mineral salt composition is less critical provided the essential nutrients for plant cell growth are included. Organic nitrogen supplied as casein hydrolysates may be beneficial. Glutamine can be used as an alternative source of organic nitrogen (Bui-Dang-Ha and MacKenzie, 1973).

The conditions developed for isolating and culturing protoplasts from shoot tips of pea (*Pisum sativum* L.) are shown in Table 1 and Figure 1. The shoot tips were taken from three day old seedlings germinated in the dark at 26°C. Protoplasts can be isolated from older seedlings, but they fail to survive in culture. Pea protoplast viability appeared to depend upon thorough removal of enzyme, which was achieved by four washings.

Fig. 1. Protoplast isolation, division and agglutination.
(a) Protoplast isolation from pea shoot tips. (b) Division of cells regenerated from bean leaf protoplasts. (c) Protoplasts from pea leaf mesophyll and *Vicia hajastana* cultured cells agglutinated by PEG. (d) Electron microscopic section of the agglutinated protoplasts shown in (c).

TABLE 1

Conditions for isolation of

protoplasts from shoot tips of pea (*Pisum sativum*)

	Isolation Conditions - two steps	
	I (1 hr)	II (4-5 hrs)
Enzymes		
Cellulase	"Driselase" 1%	1%
Cellulase	"Onozuka" 1%	0.5%
Pectinase	"Sigma" 0.5%	0.5%
Hemicellulase	"Rhozyme" 0.5%	
Sorbitol	0.25 M	0.25 M
Mannitol	0.25 M	0.25 M
$CaH_4(PO_4)_2 \cdot H_2O$	1.0 mM	1.0 mM
$CaCl_2 \cdot 2H_2O$	5.0 mM	5.0 mM
Osmolality	0.5	0.5
pH	6.2	6.2

Growing protoplasts in medium droplets of 50-150 μl generally has been more successful than on agar. Droplets with about 10^4 protoplasts per ml were placed in plastic petri dishes and incubated in humidified chambers at 25-28°C. Cell wall formation in the pea protoplasts occurred within 24 hrs and the first divisions were observed within 72 hrs. Glutamine at 2 mM increased the number of dividing protoplasts. The cells forming within six weeks were transferred to nutrient agar media and produced callus (Table 2). Since organogenesis has been achieved in pea shoot tip callus, it may also be possible to induce differentiation in the callus from the protoplasts (Gamborg *et al.*, 1974).

CELL REGENERATION AND DIVISION.

Protoplasts isolated from a variety of plant sources have been cultured and cell division induced (Fig 1). Some examples are shown in Table 3.

The newly formed cell walls appear to be similar in composition to those of the original cells (Albersheim, private communication). The details of mitosis resemble in every respect that of normal plant cells (Fowke *et al.*, 1974b).

TABLE 2

Culture media for pea shoot tip protoplasts

B5 mineral salts and vitamins[*]

	Protoplast in liquid droplets	Callus on agar plates
Sorbitol[**]	0.19 M	
Mannitol	0.19 M	
Glucose	28 mM	0.1 M
Ribose	25 mg/L	
Sucrose		20 g/L
L-glutamine	2 mM	2 mM
$CaCl_2 \cdot 2H_2O$	6.8 mM	
$CaH_4(PO_4)_2 \cdot H_2O$	0.5 mM	
Kinetin	5 μM	5 μM
2,4-D	1 μM	1 μM
NAA	0.1 μM	1 μM
pH	6.2	5.8
Agar		0.6%

[*] Gamborg *et al.*, 1968

[**] Sorbitol and mannitol may be replaced by 0.3 M glucose

TABLE 3

List of protoplasts in which cell regeneration and division has been observed

Plant	Cell Origin
Bromus inermis	culture
Ammi visnaga	culture
Daucus carota	culture, leaf
Glycine max	culture
Phaseolus vulgaris	culture, leaf
Pisum sativum	culture, leaf
Pisum sativum	shoot apex
Brassica napus	leaf
Vicia hajastana	culture
Vigne sinensis	leaf
Cicer arietinum	leaf
Linum usitatissimum	leaf
Melilotus alba	leaf
Medicago sativa	leaf

186

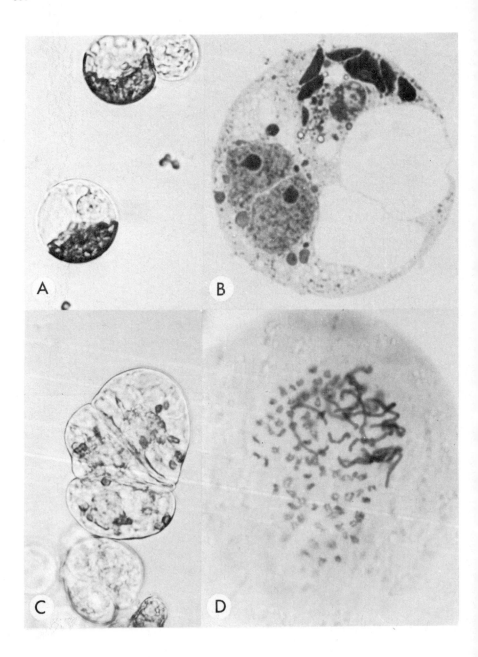

PROTOPLAST FUSION.

Protoplasts from cell cultures fuse spontaneously during isolation (Miller *et al.*, 1971; Reinert and Hellmann, 1971; Motoyoshi, 1971). Homokaryons of two and up to 100 nuclei have been observed. Heteroplasmic fusions must be induced. The spherical shape of protoplasts provides for a very small area of contact. The first step in induced fusion must therefore involve a means of agglutinating and increasing the area of membrane contact. Several agents have been reported to improve heteroplasmic fusion of protoplasts. High concentrations of nitrate were employed by several groups (Power *et al.*, 1970; Carlson *et al.*, 1972; Potrykus, 1971). Lysozyme appeared to be somewhat effective (Potrykus, 1973). Osmotic shock with centrifugation was also used, but the fusion frequency with any of these procedures was very low (Eriksson, 1970; Keller *et al.*, 1973), and there was no evidence presented to assess the rate of protoplast survival and division after these treatments.

Combined conditions of alkaline pH and high calcium salt concentration proved effective for fusion of protoplasts of different tobacco species (Keller and Melchers, 1973).

Hartmann *et al.* (1973) achieved efficient protoplast agglutination by using antiserum. The protoplasts remained viable and divided, but fusion was not achieved.

PEG-*INDUCED FUSION.*

The polyethylene glycol (PEG) procedure introduced recently (Kao and Michayluk, 1974; Constabel and Kao, 1974) effectively agglutinate and facilitate intergeneric fusion of protoplasts. Solutions of about 50% of PEG (M.W. 1540) added to protoplasts of two plant sources suspended in a droplet cause immediate agglutination (Fig 1). After 45 minutes the PEG is slowly diluted and a percentage of the protoplasts fuse and assume the normal spherical shape (Fig 2).

Fig. 2. Intergeneric protoplast fusion.
(a) Fusion product (heterokaryocyte) of protoplasts from barley mesophyll and soybean cultured cells. (b) Heterokaryocyte of protoplasts from pea leaf mesophyll and *vicia hajastana* cultured cells obtained by PEG-induced fusion. (c) Cells obtained from heterokaryocytes of soybean-barley. (d) Heterokaryocyte from soybean and *Vicia hajastana* in mitosis.

The sequency of heteroplasmic fusion (heterokaryon formation) vary
from 15-30% of the surviving protoplasts. The high rate of fusion
depends upon the quality of the protoplasts and the use of a high
concentration of PEG. The presence of calcium salts and a pH which
is alkaline may enhance the fusion process (Kao et al., 1974).

Apart from inducing fusion consistently and efficiently the PEG
treatment usually has no significant deleterious effect on the protoplasts. The compound may in fact enhance divison (see also Wallin
and Eriksson, 1973). There are no observable biological limitations
which prevent fusion of protoplasts from plants of different genera
and families (Table 4). The tissue origin also is inconsequential
(Kao and Michayluk, 1974; Constabel and Kao, 1974; Kartha et al.,
1974b; Kao et al., in press).

TABLE 4

List of plants in which heterokaryocyte formation and
division has been observed

Source of Protoplasts

Cell Culture		Leaf
Soybean (Glycine max)	x	Barley (Hordeum vulgare)
Soybean (Glycine max)	x	Corn (Zea mays)
Soybean (Glycine max)	x	Rapeseed (Brassica napus)
Soybean (Glycine max)	x	Pea (Pisum sativum)
Carrot (Daucus carota)	x	Pea (Pisum sativum)
Sweet clover (Melilotus sp.)	x	Rapeseed (Brassica napus)

HETEROKARYOCYTE SELECTION AND DIVISION.

The heterokaryocytes are recognizable by the presence of green
chloroplasts from the leaf protoplasts as well as cytoplasmic markers
from the cultured cells (Kao and Michayluk, 1974). The unfused leaf
protoplasts deteriorate because the culture conditions by design are
suitable for only one of the protoplast species. Up to 10% of the
heterokaryocytes divide within 48-72 hrs, and some form cell clusters
(Fig 2, Table 4). There is evidence of synchronous mitosis and the
presence of the chromosomes of both parental species in the hybrid
cells (Kao et al., in press; Fig 2).

The present recognition system for selecting heterokaryocytes has
been useful. However, it will be necessary to develop more efficient

selection procedures. Acceptable systems may include the use of specific pathogenic toxins (Strange, 1972), or employing cell lines with resistance to antimetabolites or herbicides (Chaleff and Carlson, 1974; Ohyama, 1974). Metabolic mutants of plant cells are difficult to obtain and may be leaky (Chaleff and Carlson, 1974), which makes them unsuitable.

The heterokaryocytes apparently form a cell wall and can be induced to divide. The chromosomes from both of the parental species can be observed during mitosis (Fig 2) but there is no information on the relative distribution in the cell progeny.

TRANSFORMATION.

Investigations on the uptake of DNA and on transformation related to phenotypic expression has been reviewed recently (Holl *et al.*, 1974; Johnson and Grierson, 1974). Plant protoplasts take up bacterial DNA relatively rapidly (Ohyama *et al.*, 1973). The process of uptake and the fate of exogenous DNA in recipient cells has not been elucidated. The present evidence indicates that bacterial DNA after uptake appears inside the protoplasts as the double stranded form. Degradation can be observed, but it could be expected that some of the DNA becomes stabilized or intergrated into the host genome (Ledoux *et al.*, 1971; Gahan *et al.*, 1973). The critical assessment of the use of isolated DNA for achieving genetic transformation in plants is limited by the availability of specific biological markers. Although protoplasts may be convenient plant materials for transformation experiments, their usefulness for the present is restricted because procedures are not generally available for inducing cell division and plant regeneration. When protoplasts are employed it is also essential to have a biological marker system which selects the transformed cells.

The capability of plant cells to utilize a new carbon source has potential as a selection system. Cultured plant cells grow predominantly on sucrose, glucose or fructose. The same applies to a soybean suspension culture in our laboratory. Other carbohydrates such as mannitol or lactose were ineffective. Protoplasts from the soybean cells were allowed to take up DNA from *Azotobacter vinelandii*. The cells regenerated from the protoplasts, divided and were plated on nutrient agar with mannitol + lactose as the carbon sources. After several months a number of cell colonies were observed on the agar. The cells have been subcultured for one year and retained their ability to grow on the new carbon sources.

Transformation experiments can also be pursued with whole plants or plant organs as the recipient materials (Ledoux *et al.*, 1971; Hess, 1972). These systems have the advantage that complete plant formation is ensured (Holl *et al.*, 1974).

MORPHOGENESIS.

Somatic plant cells may differentiate and regenerate complete plants (Street, 1969). The process can occur by embryogenesis which does not require exogenous hormones (Halparin, 1970; Grambow *et al.*, 1972). In most cases, however, organogenesis must be induced by cytokinins and auxins. There are numerous examples of plant regeneration from callus, but with few exceptions, the techniques have not been established as a routine method for mass production.

Shoot initiation is induced by cytokinins in most plants, but not in cereals and grasses. The conditions for successful plant regeneration must be determined separately for different plant tissues. In order to grow plants from cells after genetic modifications it is therefore imperative to determine the necessary conditions and requirements for inducing organogenesis in particular cell lines. It must also be demonstrated that cells regenerated from protoplasts can be induced to form plants (Fig 3). That has been shown for carrot (Grambow *et al.*, 1972), brome grass (Kao *et al.*, 1973), rapeseed (Kartha *et al.*, 1974a), tobacco (Takebe *et al.*, 1971), petunia (Durand *et al.*, 1973) and asparagus (Bui-Dang-Ha and MacKenzie, 1973). As hybrid and genetically modified cells become available it should be possible to induce morphogenesis and plant development.

DISCUSSION

The recent progress in protoplast and tissue culture technology demonstrates the feasibility of producing intergeneric cell hybrids. The natural barriers which prevent sexual crosses between plant genera are not apparent in the fusion and subsequent growth of the hybrid cells.

With present techniques the cell colonies may consist of 10-100 cells. They are identified by the presence of chloroplasts. The techniques do not provide a means of selecting and isolating the hybrid cell colonies. Selection systems must be developed which will result in the isolation of the hybrids, and deletion of parental cells. This may be achieved by mechanical means or through the use of specific toxins, or other selective chemicals. The production of hybrid

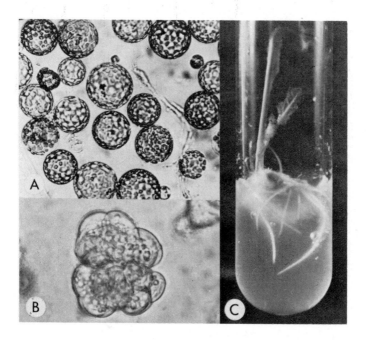

Fig. 3. Division and plant regeneration from protoplasts of rapeseed
         leaf.
         (a) Protoplasts from rapeseed mesophyll. (b) Callus from
         rapeseed protoplasts. (c) Rapeseed plantlet obtained from
         a mesophyll protoplast.

plants is handicapped by the lack of information on the biochemical
events leading to division and to differentiation in higher plants.
The capability for totipotency in plant cells is generally accepted,
but a reproducible method for inducing the process is lacking.
Except for very few species, any attempts at present to induce division and differentiation in plant protoplasts is therefore largely
empirical.

   The testing and development of genetic transfer methods also could
be facilitated by having plant materials of mutants and of isogenic
lines which differ by a single gene with respect to a measurable property (Holl et al., 1974). With the technology already available
with plants of economic importance, it is possible to direct the
methodology into applied problems (Nickell and Heinz, 1973). Methods
for producing wide crosses are being sought as an aid in plant breed-

ing and crop improvement (Borlaug, 1971). The formation of intergeneric hybrids through somatic hybridization provides plant material
with greater genetic variability than is available at present. Such
materials may facilitate plant improvement directed towards raising
protein quality, disease resistance and tolerance to extreme environmental conditions. Current objectives include plans to obtain plants
with greater photosynthetic efficiency, and to equip non-leguminous
plants with the capability to fix nitrogen from the air. The methods
of somatic cell hybridization makes it possible to investigate such
objectives (Holl et al., 1974). The techniques are also potential
aids in genetic analyses. Cell hybrids of mammalian cells are used extensively and mapping of human chromosomes has been greatly aided.
The scope with somatic cell hybridization in plants is immensely
greater because of the possibility of obtaining organs and complete
plants from the hybrid cells.

ACKNOWLEDGEMENTS

We appreciate the kindness of Dr. L. Pelcher and Dr. K. Kartha of
this laboratory, and Dr. L. Fowke, Department of Biology, University
of Saskatchewan in supplying some of the photographs. We thank
Mr. J. Shyluk for competent technical assistance and Mr. A. Lutzko
for preparing the photographs.

REFERENCES

Borlaug, W.E. (1971). The green revolution, peace, and humanity.
  Nobel Lecture. *Cereal Science Today*. 16, 401-411.
Bui-Dang-Ha, D., and MacKenzie, I.A. (1973). The division of proto-
  plasts from *Asparagus officinalis* L. and their growth and differ-
  entiation. *Protoplasma*. 78, 215-221.
Carlson, P.S., Smith, S.H., and Dearing, R.D. (1972). Parasexual
  interspecific plant hybridization. *Proceedings of the National
  Academy of Sciences of the United States of America*. 69, 2292-2294.
Chaleff, R.S., and Carlson, P.S. (1974). Somatic cell genetics of
  higher plants. *Annual Review Genetics* (in press).
Cocking, E.C. (1970). Virus uptake, cell wall regeneration and virus
  multiplication in isolated protoplasts. *International Review of
  Cytology*. 28, 89-124.
Constabel, F. (1975). Protoplast isolation and culture. In *Plant
  Tissue Culture Methods*. Edited by Gamborg, O.L. and Wetter, L.R.
  National Research Council of Canada. (In press).

Constabel, F., and Kao, K.N. (1974). Agglutination and fusion of
  plant protoplasts by polyethylene glycol. *Canadian Journal of
  Botany*. 52, 1603-1606.

Constabel, F., Kirkpatrick, J.W., and Gamborg, O.L. (1973). Callus
  formation from mesophyll protoplasts of *Pisum sativum*. *Canadian
  Journal of Botany*. 51, 2105-2106.

Driselase for protoplast preparation. (1974). Edited and produced by
  Kyowa Hakko Kogyo Co. Ltd., Tokyo, Japan.

Durand, J., Potrykus, I., and Donn, G. (1973). Plantes issues de
  protoplastes de Petunia. *Zeitschrift für Pflanzenphysiologie*.
  69, 23-34.

Eriksson, T. (1970). Isolation and fusion of plant protoplasts.
  *Colloques Internationaux du Centre National de la Recherche
  Scientifique*. 193, 297-303.

Fowke, L.C., Beck-Hansen, C.W., Gamborg, O.L., and Shyluk, J.P. (1973).
  Electron microscopic observations of cultured cells and protoplasts of *Ammi visnaga*. *American Journal of Botany*. 60, 304-312.

Fowke, L.C., Beck-Hansen, C.W., and Gamborg, O.L. (1974a). Electron
  microscopic observations of cell regeneration from cultured protoplasts of *Ammi visnaga* L. *Protoplasma*. 79, 235-248.

Fowke, L.C., Beck-Hansen, C.W., Constabel, F., and Gamborg, O.L.
  (1974b). A comparative study on the ultrastructure of cultured
  cells and protoplasts of soybean during cell division. *Protoplasma*.
  81, 189-203.

Gahan, P.B., Anker, P., and Stroun, M. (1973). An autoradiographic
  study of bacterial DNA in *Lycopersicon esculentum*. *Annals of
  Botany*. 37, 681-685.

Gamborg, O.L., Miller, R.A., and Ojima, K. (1968). Nutrient requirements of suspension cultures of soybean root cells. *Experimental
  Cell Research*. 50, 151-158.

Gamborg, O.L., and Miller, R.A. (1974). Isolation, culture and uses
  of plant protoplasts. *Canadian Journal of Botany*. 51, 1795-1799.

Gamborg, O.L., Constabel, F., and Shyluk, J.P. (1974). Organogenesis
  in callus from shoot apices of *Pisum sativum* L. *Physiological
  Plantarum*. 30, 125-128.

Gamborg, O.L., Kao, K.N., Miller, R.A., Fowke, L.C., and Constabel,
  F. (1973). Cell regeneration, division and plant development from
  protoplasts. *Colloques Internationaux du Centre National de la
  Recherche Scientifique*. 212, 155-173.

Grambow, H.J., Kao, K.N., Miller, R.A., and Gamborg, O.L. (1972).

Cell division and plant development from protoplasts of carrot cell suspension cultures. *Planta (Berlin)*. 103, 348-355.

Halparin, W. (1970). Embryos from somatic plant cells.In *Control Mechanisms in the Expression of Cellular Phenotypes*. Edited by H. Padykula, Academic Press. 169-191.

Hartmann, J.X., Kao, K.N., Gamborg, O.L., and Miller, R.A. (1973). Immunological methods for the agglutination of protoplasts from cell suspension cultures of different genera. *Planta (Berlin)*. 112, 45-56.

Hess, D. (1972). Transformationen an hoheren organismen. *Naturwissenschaften*. 59, 348-355.

Hoffmann, F., and Hess, D. (1973). Die aufname radioaktiv markierster DNA in isolierte protoplasten von *Petunia hybrida*. *Zeitschrift für Pflanzenphysiologie*. 69, 81-83.

Holl, F.B., Gamborg, O.L., Ohyama, K., and Pelcher, L. (1974). Genetic transformation in plants. In *Tissue Culture and Plant Science*. Edited by H.E. Street. Academic Press.

Johnson, C.B., and Grierson, D. (1974). The uptake and expression of DNA by plants. In *Current Advances in Plant Science*. Editor H. Smith, Maxwell, Oxford, Vol. 2, No. 9, 1-12.

Kanai, R., and Edwards, G.E. (1973). Purification of enzymatically isolated mesophyll protoplasts from C_3, C_4 and *Crassulacean* acid metabolism, plants using an aqueous Dextran-polyethylene glycol two-phase system. *Plant Physiology*. 52, 484-490.

Kao, K.N., Gamborg, O.L., Michayluk, M.R., Keller, W.A., and Miller, R.A. (1973). The effects of sugars and inorganic salts on cell regeneration and sustained division in plant protoplasts. *Colloques Internationaux de Centre National de la Recherche Scientifique*. 212, 207-213.

Kao, K.N., and Michayluk, M.R. (1974). A method for high-frequency intergeneric fusion of plant protoplasts. *Planta (Berlin)*. 115, 355-367.

Kao, K.N., Constabel, F., Michayluk, M.R., and Gamborg, O.L. Plant protoplast fusion and growth of intergeneric hybrid cells. *Planta*. (In press).

Kartha, K.K., Michayluk, M.R., Kao, K.N., and Gamborg, O.L. (1974a). Callus formation and plant regeneration from mesophyll protoplasts of rapeseed plants (*Brassica napus* L. cv. Zephyr). *Plant Science Letters*. 3, 265-271.

Kartha, K.K., Gamborg, O.L., Constabel, F., and Kao, K.N. (1974b).
Fusion of rapeseed and soybean protoplasts and subsequent division
of heterokaryocytes. *Canadian Journal of Botany*. (In press).

Keller, W.A., and Melchers, G. (1973). The effect of high pH and
calcium on tobacco leaf protoplast fusion. *Zeitschrift für Naturforschung*. 28, 737-741.

Keller, W.A., Harvey, B.H., Kao, K.N., Miller, R.A., and Gamborg,
O.L. (1973). Determination of the frequency of interspecific
protoplast fusion by differential staining. *Colloques Internationaux du Centre National de la Recherche Scientifique*. 212, 455-463.

Ledoux, L., Huart, R., and Jacobs, M. (1971). Fate of exogenous DNA
in *Arabidopsis thaliana*. Evidence for replications and preliminary
results at the biological level. In *Informative molecules in
biological systems*. Edited by L.G.H. Ledoux. North Holland Amsterdam. 159-172.

Miller, R.A., Gamborg, O.L., Keller, W.A., and Kao, K.N. (1971).
Fusion and division of nuclei in multinucleated soybean protoplasts.
Canadian Journal of Genetics and Cytology. 13, 347-353.

Motoyoshi, F. (1971). Protoplasts isolated from callus cells of
maize endosperm. *Experimental Cell Research*. 68, 452-456.

Nagata, T., and Takebe, I. (1971). Plating of isolated tobacco mesophyll protoplasts on agar medium. *Planta (Berlin)*. 99, 12-20.

Nickell, L.G., and Heinz, P.J. (1973). Potential of cell and tissue
culture techniques as aids in economic plant improvement. In
Genes, Enzymes and Populations. Edited by A.M. Srb. Plenum, New
York. 109-128.

Ohyama, K. (1974). Properties of 5-bromodeoxyuridine-resistant lines
of higher plant cells in liquid culture. *Experimental Cell
Research*. (In press).

Ohyama, K., Gamborg, O.L., and Miller, R.A. (1972). Uptake of exogenous DNA by plant protoplasts. *Canadian Journal of Botany*. 50,
2077-2080.

Ohyama, K., Gamborg, O.L., Shyluk, J.P., and Miller, R.A. (1973).
Studies on transformation. Uptake of exogenous DNA by plant protoplasts. *Colloques Internationaux du Centre National de la Recherche Scientifique*. 212, 423-428.

Pelcher, L.E., Gamborg, O.L., and Kao, K.N. (1974). Bean mesophyll
protoplasts: production, culture and callus formation. *Plant
Science Letters*. 3, 107-111.

Potrykus, I. (1971). Intra- and inter-specific fusion of protoplasts

196

from petals of *Torenia baillonii* and *Torenia fournieri*. *Nature New Biology*. 231, 57-58.

Potrykus, I. (1973). Isolation, fusion and culture of protoplasts. In *Yeast, Mould and Plant Protoplasts*. Edited by J.R. Villanueva *et al.* Academic Press, New York. 319-330.

Power, J.B., Cummins, S.E., and Cocking, E.C. (1970). Fusion of isolated plant protoplasts. *Nature*. 225, 1016-1018.

Reinert, J., and Hellmann, S. (1971). Mechanism of the formation of polynuclear protoplasts from cells of higher plants. *Naturwissenschaften*. 58, 419.

Strange, R.N. (1972). Plants under attack. *Science Progress*. 60, 365-385.

Street, H.E. (1969). Growth in organized and unorganized systems. In *Plant Physiology*. Vol. VB. Edited by F.C. Steward. Academic Press, 227-327.

Takebe, I., and Otsuki, Y. (1969). Infection of tobacco mesophyll protoplasts by tobacco mosaic virus. *Proceedings of the National Academy of Science, (Washington)*. 64, 843-848.

Takebe, I., Labib, G., and Melchers, G. (1971). Regeneration of whole plants from isolated mesophyll protoplasts of tobacco. *Naturwissenschaften*. 58, 318-320.

Tempé, J. (1973). Editor. Protoplastes et fusion de cellules somatiques vegetales. *Colloques Internationaux du Centre National de la Recherche Scientifique*. No. 212.

Wallin, A., and Eriksson, T. (1973). Protoplast cultures from cell suspension of *Daucus carota*. *Physiological Plantarum*. 28, 33-39.

HIGHER PLANT CELLS AS EXPERIMENTAL ORGANISMS

R.S. CHALEFF[*] AND P.S. CARLSON[†]

*Biology Department, Brookhaven National Laboratory,
Upton, New York 11973 U.S.A.*

The ability to culture somatic cells of higher plants
under defined conditions permits direct selection for
variants among large cell populations. Cells of tobacco resistant to the methionine analogue, methionine
sulfoximine, have been selected *in vitro* and regenerated
into mature plants. All three tobacco mutants are resistant to infection by *Pseudomonas tabaci*, the agent
which causes wildfire disease. Two of these mutants
contain elevated intracellular levels of free methionine. Analogue resistance also has been used to select
for variant rice cell lines which have increased levels
of certain free amino acids and higher relative amounts
of lysine incorporated into protein.

An experimental system also is described for forcing a
symbiotic association between cultured cells of carrot
and the free-living nitrogen-fixing bacterium, *Azotobacter vinelandii*. The composite callus grows slowly
on a defined medium lacking combined nitrogen.

Genetic studies of microorganisms have achieved a highly sophisti-
cated and refined level. The relative ease with which defined mutants are isolated in these unicellular forms has permitted the elucidation of biochemical and developmental pathways. Investigations of
microorganisms also have contributed largely to our present understanding of the molecular mechanisms which mediate the expression of
genetic information. In contrast, the extent to which we understand
and are able to manipulate the genetic content of higher plants is
severely limited.

---

[*] Present address: Department of Applied Genetics, John Innes
Institute, Colney Lane, Norwich NR4 7UH, England.

[†] Present address: Department of Crop and Soil Science, Michigan
State University, East Lansing,
Michigan 48823, U.S.A.

The principles of genetic organization derived from studies of microbial organisms should not be expected to provide an adequate realization of the genetic architecture of higher eukaryotes. The cell of a higher eukaryotic organism is organized very differently and inhabits and must respond to a much different environment than cells of lower forms. However, the experimental approach of molecular genetics is still valid and may be applied to the analysis of these more complex forms. The very differences in biological organization which necessitate these studies are primarily responsible for the difficulty in recovering defined mutants. Bacteria and fungi have extended haploid phases and small nutrient reserves which permit the immediate phenotypic expression of genetic variation. The ability to grow large, homogeneous populations with short generation times on defined media makes possible the application of selective screens to enormous numbers of genomes. These organizational features, which have nearly restricted the molecular genetical approach to microbes, are now becoming available to higher plants. Cells of many plant species may be cultured under defined conditions (Street, 1973); techniques exist for obtaining haploid cell lines (Chase, 1969; Nitsch, 1972; Sunderland, 1973); and whole plants may be differentiated from cultured cells (Vasil and Vasil, 1972; Reinert, 1973). With the ability to manipulate experimentally higher plant cells as microorganisms, it should be possible to dissect the functioning of these more complex forms. This understanding could then be applied to effect agronomically beneficial changes in important crop species.

MUTANTS WHICH OVERPRODUCE AMINO ACIDS

Several experiments have illustrated the feasibility of applying microbial selective systems to cultured cells of higher plants to achieve directed modification of plant genomes. In one experiment, mutants of *Nicotiana tabacum* resistant to the wildfire disease were recovered (Carlson, 1973). This disease of tobacco is caused by a bacterial pathogen, *Pseudomonas tabaci*, which produces a toxin structurally similar to methionine (Braun, 1955). Resistant tobacco cells were selected among a mutagenized haploid cell population for their ability to grow in a medium containing an inhibitory concentration of methionine sulfoximine. Methionine sulfoximine is an analogue of the wildfire toxin which elicits the formation of the same characteristic chlorotic halos on tobacco leaves as does the natural toxin (Braun, 1955). Three homozygous diploid plants which were regenerated from these methionine sulfoximine-resistant calluses are less susceptible

Table 1

Concentrations of certain free amino acids in
tobacco leaves (nmol per gram fresh weight)

Strains	Methionine	Glycine	Alanine	Proline
a) Homozygous				
Havana Wisconsin 38	0.4±0.2	1.3±0.3	1.8±0.3	0.3±0.1
Mutant 1	0.3±0.2	1.4±0.3	1.7±0.5	0.4±0.2
Mutant 2	1.9±0.5	1.7±0.5	2.0±0.4	0.5±0.2
Mutant 3	2.4±0.6	1.2±0.2	1.5±0.3	0.4±0.2
b) Heterozygous				
Mutant 1/+	0.5±0.2	1.6±0.4	2.1±0.5	0.3±0.2
Mutant 2/+	0.9±0.4	1.5±0.3	1.7±0.4	0.5±0.2
Mutant 3/+	1.5±0.7	1.1±0.3	1.9±0.5	0.3±0.2

Fully expanded young leaves were deveined and homogenized at room
temperature. An equal volume of 10% trichloroacetic acid was added
to the homogenate. The acid soluble supernatant was applied to an
automated amino acid analyzer.

than the parent plant to the pathogenic effects of bacterial infection.
Resistance appears to be genetically complex in one mutant plant which
contains wild-type levels of free methionine. The resistance phenotype of the two other plants is transmitted in crosses as a single
semidominant locus. The intracellular concentration of free methionine in these two homozygous plants is five times higher than in the
wild-type (Table 1a). That the amount of free methionine in the heterozygous strain (Table 1b) is half of that observed in the homozygous
mutant plant is consistent with the semidominant phenotype of the
mutations. The recovery of tobacco plants resistant to wildfire disease reveals the potentiality of selection for toxin resistance *in
vitro* as a generalized procedure for obtaining disease resistant
varieties. These results also demonstrate that mutants of higher
plants in which the regulation of amino acid biosynthesis is altered
may be recovered *in vitro*. This latter conclusion has been extended
by additional research. Widholm (1972a, b) has isolated cell lines
of tobacco and carrot which are capable of growth in the presence of
a normally inhibitory concentration of 5-methyl tryptophan. Crude
extracts of the resistant cell lines contain a species of anthranilate
synthetase which is less sensitive to feedback inhibition by trypto-

phan and 5-methyl tryptophan than is the wild-type enzyme. Endogenous levels of free tryptophan in resistant cell lines of tobacco
and carrot are fifteen times and 27 times higher, respectively, than
the wild-type levels. Thus, it is evident that in higher plants the
endogenous concentration of a specific metabolite may be increased by
selecting for resistance to a structural analogue of that metabolite.
It follows, therefore, that by using analogues of amino acids essential to human nutrition, variants of crop species which produce elevated levels of these compounds may be induced and selected in culture.

EXPERIMENTS WITH CULTURED RICE TISSUE

The seed endosperm protein of rice is of inferior nutritional quality because of relative insufficiencies of lysine, threonine, and isoleucine. In higher plants these three amino acids and methionine,
which is also nutritionally essential, are derived from aspartate by
the biosynthetic pathway outlined in Figure 1 (Miflin, 1973).

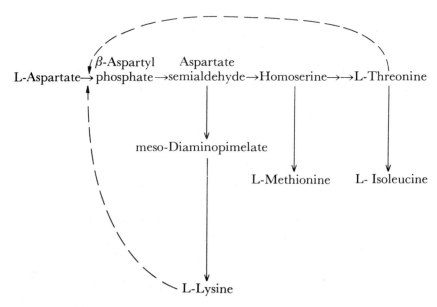

Fig 1 Biosynthetic pathway of the aspartate family of amino acids.
Dashed lines denote hypothetical patterns of feedback inhibition.

The activity of the first enzyme, aspartokinase, appears to be regulated through feedback inhibition primarily by lysine and to a lesser degree by threonine (Bryan et al., 1970; Henke et al., 1974; Furuhashi and Yatazawa, 1970). Aspartokinase activity in cell-free extracts of rice shoots is also sensitive to feedback inhibition by the lysine analogue S-(β-aminoethyl)-cysteine (SAEC) (Halsall et al., 1972). These observations suggested that resistance to SAEC could be utilized to select for rice mutants in which lysine biosynthesis is released from regulatory control. Furthermore, it was anticipated that a mutation conferring analogue resistance due to synthesis of a feedback insensitive enzyme would be semidominant. Thus, the experiment could be carried out with cultured diploid cells by a procedure similar to that used by Widholm (1972a, b).

Rice callus cultures were obtained from seed by the method of Yamada et al. (1967) and were maintained either in liquid or on solid medium containing the mineral salts and vitamin concentrations described by Linsmaier and Skoog (1965) supplemented with 3% sucrose and 0.5 mg 2,4-D, 5.0 mg IAA and 0.3 mg kinetin per liter. Suspensions of small cell aggregates were mutagenized with 1% ethylmethanesulfonate for 1 hr and washed and resuspended in fresh medium. The cultures were incubated for an additional 10 days to ensure genetic segregation and phenotypic expression of any mutational event. The tissue was then plated on the basic medium supplemented with 2 mM SAEC, a concentration of the analogue which kills wild-type rice cells. Resistant clones were selected as they appeared over a two month period. Three variant cell lines were recovered by this procedure which survive and grow very slowly in the presence of the analogue. The intracellular concentration of free amino acids in these three variant cell lines and in the wild-type are presented in Table 2. The three variant lines contain elevated levels of three of the amino acids derived from aspartate as well as leucine, valine, tyrosine, and alanine. Threonine was not resolved by the chromatographic system which used sodium citrate buffer. The largest increases, which were observed in the levels of free isoleucine and leucine, are not as great as the tryptophan increases reported by Widholm (1972a, b) in 5-methyl-tryptophan-resistant lines of tobacco and carrot. However, these increases are comparable with the specific increase in free methionine observed in methionine sulfoximine-resistant tobacco leaves (Table 1). If the variant rice cell lines are presumed to be heterozygous, since they were isolated from diploid tissue, any changes

Table 2

Free amino acid pools of rice tissue cultures

(expressed as nmols/mg dry weight)

Amino acid	Wild-type	EMS-5	EMS-6	EMS-8
Lysine	7.67	14.07	14.84	19.47
Histidine	10.76	6.21	5.97	6.96
Arginine	5.90	5.97	8.74	10.99
Glycine	6.89	11.17	17.27	17.89
Alanine	1.82	22.35	8.40	23.75
Valine	2.84	8.18	12.02	14.25
Methionine	0.71	1.86	1.80	2.06
Isoleucine	0.79	3.10	6.44	7.04
Leucine	0.80	3.88	6.26	6.67
Tyrosine	0.97	2.87	2.74	3.64
Phenylalanine	10.22	6.91	5.41	7.51
Total nmol per mg dry wt	104.10	182.32	149.31	193.06

Free amino acids were extracted from lyophilized callus tissue in H_2O at $85°C$. Samples were concentrated to dryness, redissolved in sodium citrate buffer, and applied to an automated amino acid analyzer.

in the free amino acid concentrations should be comparable with those in the heterozygous mutant strains of tobacco (Table 1b). It is seen that free lysine, methionine, and valine are increased by the same factor as is methionine in the heterozygous tobacco mutant. The increase in free lysine is not sufficient to explain the increase in total lysine (approximately 30 nmoles per mg dry weight) observed in the variant cell lines. The data in Table 3 were obtained from acid hydrolysis of callus tissue and represent the sum of free and protein-bound amino acids. These data are expressed as mole percent of total amino acid content and, therefore, reflect protein quality. The percent change from the wild-type values are given in brackets. The increases in the relative levels of lysine, isoleucine, leucine, and valine evidence the improved quality of total cell protein in the three variant lines. That analogue resistance has selected for strains altered not only in endogenous concentrations of free amino acids, but also in the relative amount of each amino acid incorporated into protein, is an interesting and unexpected result.

Table 3

Total amino acid composition of rice tissue cultures
(expressed as percent moles of total)

Amino acid	Wild-type	EMS-5		EMS-6		EMS-8	
Lysine	5.19	6.54	(+26.0)	5.86	(+12.9)	6.26	(+20.6)
Histidine	2.49	2.13	(-14.5)	2.05	(-17.7)	2.12	(-14.9)
Arginine	4.19	4.69	(+11.9)	4.72	(+12.6)	4.78	(+14.1)
Aspartate	11.37	10.55	(- 7.2)	9.85	(-13.4)	9.91	(-12.8)
Threonine	5.26	5.02	(- 4.6)	5.13	(- 2.5)	5.04	(- 4.6)
Serine	6.91	6.79	(- 1.7)	6.65	(- 3.8)	6.73	(- 2.6)
Glutamate	8.86	11.15	(+25.8)	9.91	(+11.9)	10.12	(+14.2)
Proline	5.49	5.40	(- 1.6)	5.93	(+ 8.0)	5.58	(+ 1.6)
Glycine	17.50	12.05	(-31.1)	13.69	(-21.8)	13.12	(-25.0)
Alanine	9.33	10.15	(+ 8.8)	9.70	(+ 4.0)	10.20	(+ 9.3)
Valine	6.37	6.94	(+ 8.9)	7.42	(+16.5)	7.23	(+13.5)
Isoleucine	3.97	4.41	(+11.1)	4.44	(+11.8)	4.45	(+12.1)
Leucine	6.79	7.87	(+15.9)	7.88	(+16.1)	7.81	(+15.0)
Tyrosine	2.21	2.56	(+15.8)	2.89	(+30.8)	2.80	(+26.7)
Phenylalanine	4.08	3.75	(- 8.1)	3.91	(- 4.2)	3.85	(- 5.6)
				Percent protein			
	22.9	25.8	(+12.9)	23.2	(+ 1.6)	28.3	(+23.7)

Lyophilized callus tissue was hydrolyzed in 6 N HCl for 22 hrs
according to the procedure of Moore and Stein (1963).

There is also some indication that the total amount of protein is
higher in EMS-5 and EMS-8 than in the cell line from which these
variants were isolated (Table 3). However, these differences in protein content should be interpreted with caution, since they were
somewhat variable and the extent of dehydration may not have been constant for the lyophilized samples upon which the dry weight determination was based. Differences were also noted in the amino acid com-
positions between several wild-type rice cell lines, but not within
a given cell line. This variability may be due to physiological
differences making valid only a comparison between analyses of variant cell lines and the original untreated cell line from which they
were derived. Analogue resistance selects merely for expression of a
specific phenotype. The mechanisms which effect this phenotype will
remain unknown until mature plants are differentiated and subjected
to genetic analysis.

The growth responses of the wild-type and variant cell lines to several mixtures of exogenous amino acids should provide further information concerning the nature of the phenotypic differences and also enable selective conditions for the isolation of additional variants to be optimized. The results in Table 4 represent the growth of the wild-type and three variant cell lines following a 22 day incubation period on a culture medium supplemented with combinations of lysine, threonine, methionine, and isoleucine. A mixture of lysine and threonine inhibits growth of rice callus tissue presumably by feedback inhibition of aspartokinase and resultant starvation for methionine (Furuhashi and Yatazawa, 1970; Fig 1). EMS-6 is insensitive to this effect. EMS-8 is less sensitive than wild-type to inhibition by lysine alone and growth of EMS-6 is unaffected by the addition of 4 mM lysine to the medium. Whereas 2 mM lysine partially diminishes the inhibitory effect of 2 mM SAEC on the wild-type cell line, the growth response of EMS-5 to the analogue is unaltered by lysine. In general, growth of EMS-5 is affected more adversely by exogenous amino acids than is that of the wild-type tissue. This is particularly evident in the response of these cell lines to 2 mM isoleucine. The differences in growth responses of the three variants to the mixtures of amino acids make it probable that independent events are responsible for the observed phenotypes. Although there is no apparent difference between the variant and wild-type cell lines after 22 days in the presence of 2 mM SAEC, a differential growth response is evident after 6-8 weeks which permits variants to be selected in culture. The analogue-resistant phenotype of the three variant cell lines has been expressed stably *in vitro* for more than one year, even following extensive periods of subculture in the absence of the analogue. The results in Table 4 indicate that variants similar to EMS-6 could be recovered efficiently on a medium supplemented with lysine and threonine. These experiments are now in progress.

It is apparent from the results reported that it is possible to select in culture for variant cell lines of rice in which the quality and perhaps the quantity of cell protein are altered favourably. It is not known whether these changes will be reflected in the seed endosperm protein. This question is crucial and can be answered only by studying plants obtained from variant cell lines selected *in vitro*. Unfortunately, the variants isolated in this work were recovered from rice tissue which had been maintained in culture for too long a period and no longer retained its morphogenetic capacity (Nishi *et al.*, 1968).

Table 4

Growth responses of wild-type and variant cell lines
to exogenous amino acids

Amino acids	Wild-type	EMS-5	EMS-6	EMS-8
2 mM Lys	17.5	10.4	147.1	91.3
2 mM SAEC	1.1	0.5	1.7	0
2 mM SAEC+2 mM Lys	15.5	0.1	50.5	17.1
4 mM Lys	11.5	0.1	105.3	57.5
2 mM Ile	61.8	0	50.5	6.7
2 mM Thr	18.6	9.9	96.6	0
2 mM Met	8.2	0.4	20.5	12.8
2 mM Thr+2 mM Met	17.7	0	29.0	5.9
2 mM Lys+2 mM Thr	0	0.4	112.5	0
2 mM Lys+2 mM Thr+2 mM Met	43.5	0	95.2	26.8
2 mM Met+2 mM Thr+2 mM Ile	13.6	0	39.2	0
2 mM Lys+2 mM Thr+2 mM Ile	2.2	0	49.6	9.0
2 mM Lys+2 mM Thr+ 2 mM Met+2 mM Ile	20.8	0.2	54.3	4.4
No supplements	100 (15.85 mg)	18.71 mg	8.87 mg	10.93 mg

Initial inoculum = 0.85 mg dry weight; time period = 22 days;
numbers represent percent of growth (increase in dry weight)
on supplemented medium.

However, it is hoped that these experiments have served to define conditions which will permit the selection of similar variants in totipotent tissue of rice and other food crops.

PLANT-BACTERIAL ASSOCIATIONS

Symbiotic associations with microorganisms enable many plant
species to utilize molecular nitrogen (Stewart, 1966; Bond, 1968).
In several laboratories *in vitro* systems are being developed for the
study of these natural associations. The reduction of acetylene to
ethylene, a sensitive assay of bacterial nitrogenase activity (Hardy
et al., 1968), has been reported in *Rhizobium*-infected soybean cultures (Holsten *et al.*, 1971; Phillips, 1974; Child and LaRue, 1974).

In one instance the bacteria were shown to be located within the cytoplasm of the infected plant cell (Holsten *et al.*, 1971). The uptake
of *Rhizobium* cells by isolated protoplasts of *Pisum sativum* has also
been accomplished (Davey and Cocking, 1972).

We have attempted to define an experimental system for investigating the possibility of extending the nitrogen-fixing symbiosis to
additional crop species. Tissue culture techniques have been used to
force an association between the free-living nitrogen-fixing bacterium,
Azotobacter vinelandii, and cells of carrot, *Daucus carota*. This
system is based upon the establishment of a condition of mutual de-
pendency between an auxotrophic strain of *Azotobacter* and carrot cells
cultured on a medium lacking combined nitrogen. By selecting a free-
living nitrogen-fixing bacterial species for these experiments, we
hoped to bypass the complex interactions which have been evolved in
natural symbioses.

Carrot cells were grown on a Linsmaier and Skoog (1965) medium
supplemented with 4% sucrose and containing 3 mg indoleacetic acid
and 3 mg 6(γ,γ-dimethylallylamino)-purine (2iP) per liter. NH_4NO_3
and KNO_3 were omitted from N-free medium. Medium which lacked
NH_4NO_3 and contained 0.19 g KNO_3 per liter is referred to as low N
medium.

In order to form the *Azotobacter*-plant cell association, carrot
cell suspensions growing logarithmically in a liquid medium containing standard nitrogen levels were inoculated with cells of the adenine-requiring strain of *Azotobacter vinelandii* to a final concentration of 10^6 bacterial cells per ml. After 12 days the mixed cultures
were washed and resuspended in N-free carrot medium. The cultures
were incubated for an additional two weeks after which they were
plated in N-free carrot medium solidified with 1% Noble agar. The
plates were incubated at $23^\circ C$ in a 16 hr light/8 hr dark cycle. The
rare and slowly growing colonies were selected as they appeared three
to six months after plating and were transferred to either N-free or
low N medium. The cultures were maintained on the same media and
were transferred approximately every six weeks. The basis of the
ability of the *Azotobacter*-infected tissue to grow in the absence of
combined nitrogen was analyzed by several independent methods.

GROWTH STUDIES.

Carrot cells are unable to survive *in vitro* in the absence of
combined nitrogen. When 1.9 mM KNO_3 is supplied as the sole source

of combined nitrogen, the carrot cell mass survives over a four month period, but does not proliferate. Calluses derived from *Azotobacter*-inoculated carrot cell suspensions increase in fresh weight on these media. This increase is significantly greater on the low nitrogen medium than on an N-free medium, indicating that nitrogen is limiting to growth. On medium containing 20.6 mM NH_4NO_3 and 18.8 mM KNO_3 (standard concentrations), no difference in growth rates is observed between uninoculated calluses and inoculated calluses which are capable of growth on N-free medium (Fig 2). Although electron microscopy does not reveal cell divisions in the *Azotobacter*-inoculated callus, thin cell walls are observed with sufficient frequency to suggest that the carrot tissue maintains a moderate rate of mitotic activity.

REISOLATION OF Azotobacter CELLS.

Azotobacter-infected calluses capable of growth on N-free or low N medium were suspended in T broth and in Burk's minimal and adenine-supplemented nitrogen-free media. After several days, turbidity developed only in the adenine-supplemented Burk's medium. This suspension was plated and identified as the original *Azotobacter* adenine auxotroph with which the calluses had been inoculated. The absence of growth in T broth, which supports growth of a prototrophic *Azotobacter* strain, but not of the adenine-requiring strain, after a three week period indicates that the calluses contain no contaminating microorganism. None of these three media became turbid when inoculated with control carrot callus.

NITROGENASE ACTIVITY.

Azotobacter-containing carrot tissue capable of growth on N-free medium evolves significantly more ethylene in the acetylene-reduction assay for nitrogenase than does the uninoculated control callus (Table 5). The increased ethylene production by the composite callus is observed only in the presence of acetylene and, therefore, cannot be due to endogenous ethylene synthesis by the plant tissue.

PENICILLIN TREATMENT.

The ability of the inoculated callus to grow on N-free medium is destroyed by the addition of penicillin G at a concentration which is known to kill *Azotobacter* cells (50 µg/ml). Growth of either inoculated or uninoculated calluses on medium containing normal levels of combined nitrogen is not inhibited by the addition of penicillin. These observations indicate that the ability of the inoculated callus to grow on N-free medium is dependent upon the presence of functional

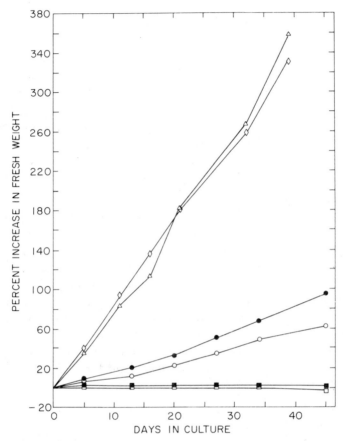

Fig 2 Growth of *Azotobacter*-infected and uninfected control carrot
tissues on media containing different levels of combined nitrogen.
Calluses were grown on low N medium for three weeks before being
transferred to the medium indicated. Control carrot tissue transferred to N-free (□), low N (■), or standard Linsmaier and
Skoog (△) medium. *Azotobacter*-containing carrot tissue transferred to N-free (o), low N (●), or standard Linsmaier and Skoog
(◊) medium.

bacterial cells in the callus mass.

ELECTRON MICROSCOPY.

Electron micrographs of inoculated carrot callus grown on low N
medium clearly establish the presence of bacterial cells in the in-
tercellular regions of the tissue (Fig 3). *Azotobacter* cells are
also found in the medium beneath the callus. No live bacteria were

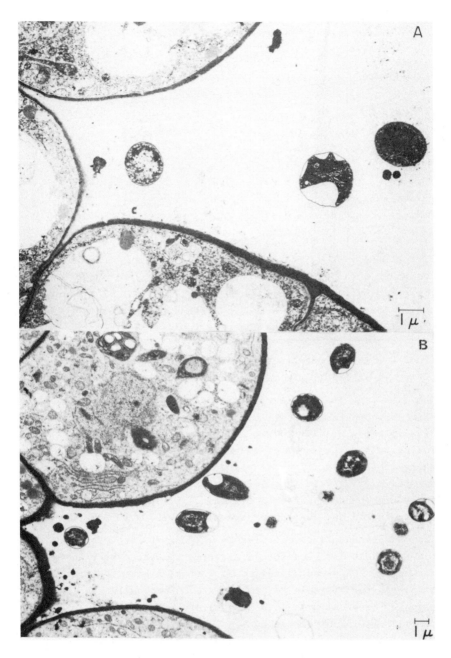

Fig 3a-b Electron micrographs of carrot-*Azotobacter* composite callus.

Table 5

Acetylene-reduction activity

Tissue	Medium	Incubation period (hr)	C_2H_2	nmol C_2H_4/ml gas
No tissue	--	24	+	0.17
Azotobacter	N free	1	+	1.91
Carrot	+N	24	-	0
Carrot	N free	24	-	0
Carrot	+N	24	+	0.31
Carrot	N free	24	+	0.30
Carrot-*Azotobacter*	N free	24	-	0
Carrot-*Azotobacter*	N free	24	+	1.84
Carrot-*Azotobacter*	N free	0	+	0.28

Approximately 50 mg fresh tissue were placed in 1 ml vials which were then injected with 0.10 ml acetylene where indicated. Reactions were stopped by the addition of 1.0 ml 0.1 N H_2SO_4. The ethylene content of samples was determined by published gas chromatographic procedures (Burris, 1972) in the laboratory of Dr. R.H. Burris.

---

observed within the carrot cells. Contaminating microorganisms were not detected in either the callus or the underlying medium.

The fine structure of the bacteria present in the cultured tissue (Fig 3a) is comparable in all essential details to that of *Azotobacter vinelandii* (Vela *et al*., 1970). The bacterial cells found within the callus contain a large number of internal vesicles near the cell periphery. These vesicles are characteristic of *Azotobacter vinelandii* cells growing at the expense of atmospheric nitrogen. Fewer vesicles are found in bacteria utilizing NH_4 and none are visible in cells growing on NO_3 (Oppenheim and Marcus, 1970).

The independent lines of evidence which have been presented suggest that a symbiosis has been forced *in vitro* between carrot and *Azotobacter* cells. Only carrot cell cultures which have been inoculated with *Azotobacter* are able to proliferate on medium lacking combined nitrogen. These cultures have been maintained for over one year on N-free medium without any reduction in growth rate. *Azotobacter*-inoculated tissue regains its dependence on an exogenous source of combined nitrogen in the presence of penicillin. Penicillin does not affect the growth of control carrot or mixed carrot-*Azotobacter* calluses on medium containing combined nitrogen. Thus, it appears that a prokaryote is providing the carrot cells with a source of reduced nitrogen. This conclusion is supported by the significantly higher levels of acetylene-reduction activity in the *Azotobacter*-inoculated calluses than in the uninfected control calluses. Finally, the presence of *Azotobacter* cells in the callus capable of growth on N-free medium was demonstrated by electron microscopy and by reisolation of the original *Azotobacter* adenine auxotroph with which the calluses had been infected.

The association between cells of carrot and *Azotobacter* was accomplished by designing a system in which the two components are able to survive only by entering into a relationship in which each complements a deficiency of the other. Presumably, carrot cells receive reduced nitrogen from *Azotobacter* cells which, in turn, depend upon the carrot cells to satisfy their auxotrophic requirement for adenine.

The experiments reported here describe an incipient system which must be refined much further before any applications may be considered. The carrot-*Azotobacter* association is not completely stable. Occasionally, sectors of a callus will lose the ability to grow on N-free medium. Other calluses become overgrown by bacteria. Because these bacteria grow in the absence of both combined nitrogen and adenine, they are assumed to be prototrophic revertants of the original *Azotobacter* strain. This latter form of instability may be controllable by using nonrevertible *Azotobacter* strains in which essential genes have been deleted. The composite callus grows very slowly on either N-free or low N medium. We are attempting to increase the flow of reduced nitrogen to the plant cells by using *Azotobacter* strains which are derepressed for nitrogenase synthesis.

An assessment of the agricultural usefulness of the carrot-*Azotobacter* association must await the regeneration of mature plants from the composite calluses. Thus far, we have been unable to accomplish this. The use of genetic markers and nutritional requirements to force a symbiosis is a generalized procedure which should be applicable to all plant species which can be grown in culture.

ACKNOWLEDGEMENTS
Research carried out at Brookhaven National Laboratory under the auspices of the U.S. Atomic Energy Commission. The authors thank Drs. L.J. Greene, R.H. Burris and M.C. Ledbetter for generously lending their advice and laboratory facilities to this project. The ex-

cellent technical assistance of Ms. B. Floyd, Mr. J. Tilley, Ms. R.
Shapanka, Mr. R. Ruffing and Mr. W. Geisbusch also contributed to
this research as did the critical and helpful discussions with Drs.
J. Polacco, D. Parke and T. Rice.

REFERENCES

Bond, G. (1968). Some biological aspects of nitrogen fixation. In
    'Recent Aspects of Nitrogen Metabolism in Plants', pp. 15-25.
    Edited by E.J. Hewitt and C.V. Cutting, New York: Academic Press.
Braun, A.C. (1955). A study on the mode of action of the wildfire
    toxin. *Phytopathology*. 45, 659-664.
Bryan, P.A., Cawley, R.D., Brunner, C.E., and Bryan, J.K. (1970).
    Isolation and characterization of a lysine-sensitive aspartokinase
    from a multicellular plant. *Biochemical and Biophysical Research
    Communications*. 41, 1211-1217.
Burris, R.H. (1972). Nitrogen-fixation-assay methods and techniques.
    In 'Methods in Enzymology'. 24, 415-431. Edited by A. San Petro,
    New York: Academic Press.
Carlson, P.S. (1973). Methionine sulfoximine-resistant mutants of to-
    bacco. *Science*. 180, 1366-1368.
Chase, S.S. (1969). Monoploids and monoploid-derivatives of maize.
    Botanical Review. 35, 117-167.
Child, J.J., and LaRue, T.A. (1974). A simple technique for the esta-
    blishment of nitrogenase in soybean callus culture. *Plant Physio-
    logy*. 53, 88-90.
Davey, M.R., and Cocking, E.C. (1972). Uptake of bacteria by isolated
    higher plant protoplasts. *Nature*. 239, 455-456.
Furuhashi, K., and Yatazawa, M. (1970). Methionine-lysine-threonine-
    isoleucine interrelationships in the amino acid nutrition of rice
    callus tissue. *Plant and Cell Physiology*. 11, 569-578.
Halsall, D., Brock, R.D., and Langridge, J.B. (1972). Selection for
    high lysine mutants. *CSIRO Division of Plant Industry Genetics
    Report*, 31-32.
Hardy, R.W.F., Holsten, R.D., Jackson, E.K., and Burns, R.C. (1968).
    The acetylene-ethylene assay for N_2 fixation: laboratory and
    field evaluation. *Plant Physiology*. 43, 1185-1207.
Henke, R.R., Wilson, K.G., and McClure, J.W. (1974). Lysine-methio-
    nine-threonine interactions in growth and development of *Mimulus
    cardinalis* seedlings. *Planta*. 116, 333-345.
Holsten, R.D., Burns, R.C., Hardy, R.W.F., and Hebert, R.R. (1971).
    Establishment of symbiosis between *Rhizobium* and plant cells

in vitro. *Nature*. 232, 173-176.

Linsmaier, E.M., and Skoog, F. (1965). Organic growth factor requirements of tobacco tissue cultures. *Physiologica Plantarum*. 18,
100-127.

Miflin, B.J. (1973). Amino acid biosynthesis and its control in plants.
In 'Biosynthesis and its Control in Plants', pp. 49-68. Edited by
B.V. Milborrow, London: Academic Press.

Moore, S., and Stein, W.H. (1963). Chromatographic determination of
amino acids by the use of automatic recording equipment. In 'Methods in Enzymology'. 6, 819-831. Edited by S.P. Colowick and N.O.
Kaplan, New York: Academic Press.

Nishi, T., Yamada, Y., and Takahashi, E. (1968). Organ redifferenti-
ation and plant restoration in rice callus. *Nature*. 219, 508-509.

Nitsch, J.P. (1972). Haploid plants from pollen. *Zeitschrift für
Pflanzenzuchtung*. 67, 3-18.

Oppenheim, J., and Marcus, L. (1970). Correlation of ultrastructure
in *Azotobacter vinelandii* with nitrogen source for growth.
Journal of Bacteriology. 101, 286-291.

Phillips, D.A. (1974). Factors affecting the reduction of acetylene by
Rhizobium-soybean cell associations *in vitro*. *Plant Physiology*.
53, 67-72.

Reinert, J. (1973). Aspects of organization-organogenesis and embryogenesis. In 'Plant Tissue and Cell Culture', pp. 338-355. Edited
by H.E. Street, Berkeley: University of California Press.

Stewart, W.D.P. (1966). 'Nitrogen Fixation in Plants', London: Athlone
Press.

Street, H.E. (1973). 'Plant Tissue and Cell Culture', Berkeley:
University of California Press.

Sunderland, N. (1973). Pollen and anther culture. In 'Plant Tissue and
Cell Culture', pp. 205-239. Edited by H.E. Street. Berkeley:
University of California Press.

Vasil, I.K., and Vasil, V. (1972). Totipotency and embryogenesis in
plant cell and tissue cultures. *In Vitro*. 8, 117-125.

Vela, G.R., Cagle, G.D., and Holmgren, P.R. (1970). Ultrastructure of
Azotobacter vinelandii. *Journal of Bacteriology*. 104, 933-939.

Widholm, J.M. (1972a). Cultured *Nicotiana tabacum* cells with an altered anthranilate synthetase which is less sensitive to feedback
inhibition. *Biochimica et Biophysica Acta*. 261, 52-58.

Widholm, J.M. (1972b). Anthranilate synthetase from 5-methyltryptophan
susceptible and resistant cultured *Daucus carota* cells. *Biochimica*

et Biophysica Acta. 279, 48-57.

Yamada, Y., Tanaka, K., and Takahashi, E. (1967). Callus induction. in rice, *Oryza sativa* L. *Proceedings of the Japan Academy.* 43, 156-160.

VIROIDS

T. O. DIENER

Plant Virology Laboratory,
Agricultural Research Service,
U.S. Department of Agriculture,
Beltsville, Maryland 20705 U.S.A.

The spindle tuber disease of potato, which was previously believed to be of viral aetiology, has been
shown to be caused by an infectious RNA of low molecular weight (*ca.* 8 x 10^4 daltons). This agent is
believed to be the first representative of a newly
recognized class of pathogens, the viroids, which
are characterized by the absence of a dormant phase
(virions) and by genomes that are much smaller than
those of known viruses. In spite of the small amount
of genetic information which viroids introduce into
their host cells, they are able to replicate and, in
some hosts, to produce disease.

The potato spindle tuber viroid has now been purified,
and some of its physical and chemical properties have
been determined.

The term "viroid" has been introduced to denote a newly recognized
class of subviral pathogens (Diener, 1971b). Presently known viroids
consist solely of a short strand of RNA with a molecular weight in
the neighbourhood of 75-100,000 Daltons. Introduction of this low-
molecular-weight RNA into susceptible hosts leads to replication of
the RNA and, in some hosts, to disease.

The first viroids came to light during efforts to purify and
characterise the agent of the potato spindle tuber disease (PSTV), a
disease which, for many years, had been assumed to be of viral
aetiology (Diener and Raymer, 1971). Diener and Raymer (1967) reported that the infectious agent of this disease is a free RNA and that
virus particles, apparently, are not present in infected tissue.
Later, sedimentation and gel electrophoretic analyses conclusively
demonstrated that the infectious RNA has a very low molecular
weight (Diener, 1971b) and that the agent, therefore, differs basic-
ally from conventional viruses.

Two additional plant diseases, chrysanthemum stunt (Diener and Lawson, 1973) and citrus exocortis (Sänger, 1972; Semancik and Weathers, 1972), are now known also to be caused by low-molecular-weight RNAs; i.e., by viroids.

The recognition of viroids as a newly identified class of pathogens raises several questions that have potentially important implications for microbiology, molecular biology, plant pathology, and veterinary and human medicine.

PROPERTIES OF VIROIDS

SEDIMENTATION PROPERTIES AND NUCLEASE SENSITIVITY.

Diener and Raymer (1967) showed that most of the infectious material in crude extracts prepared from potato or tomato leaves affected with the potato spindle tuber disease sediments in sucrose gradients at a very low rate (*ca.* 10 S). Treatment of crude extracts with phenol affected neither infectivity nor the sedimentation properties of the agent. Incubation of extracts with nucleases revealed that the agent is sensitive to ribonuclease, but not to deoxyribonuclease. In view of these findings, we proposed that the agent is a free nucleic acid.

Somewhat similar results were later reported by Singh and Bagnall (1968), who also worked with the potato spindle tuber disease, by Semancik and Weathers (1968) with citrus exocortis disease, and by Lawson (1968) with chrysanthemum stunt disease.

ABSENCE OF VIRIONS.

Although there was little doubt that the slowly sedimenting infectious material was, in each case, free RNA, the question arose as to whether this RNA exists as such *in situ* or whether it is released from conventional virus particles during extraction. A systematic study of this question led to results that are incompatible with the concept that conventional viral nucleoprotein particles exist in PSTV-infected tissue (Diener, 1971a). Furthermore, comparisons of proteins isolated from PSTV-infected tissue with those isolated from healthy tissue gave no evidence for the synthesis in infected leaves of proteins that could be construed as viral coat proteins, under conditions where coat protein of defective strains of tobacco mosaic virus could readily be demonstrated (Zaitlin and Hariharasubramanian, 1972).

SUBCELLULAR LOCATION.

Isolation of subcellular particles from PSTV-infected tissue revealed that appreciable infectivity is present only in the original tissue debris and in the fraction containing nuclei. Chloroplasts, mitochondria, ribosomes, and the "soluble" fraction contain no more than traces of infectivity. Furthermore, when chromatin was isolated from infected tissue, most infectivity was associated with it and could be extracted with phosphate buffer as free RNA (Diener, 1971a). These and other experiments suggest that, *in situ*, PSTV is associated with the nuclei and, particularly, with the chromatin of infected cells.

RECOGNITION OF LOW MOLECULAR WEIGHT OF VIROIDS.

The low sedimentation rate of PSTV is consistent with a viral genome of conventional size (> 10^6 Daltons) only if the RNA is double- or multi-stranded. Early experiments indeed indicated that the RNA might be double-stranded (Diener and Raymer, 1969), but later results showed that its chromatographic properties are not compatible with this hypothesis (Diener, 1971c).

Evidently, determination of the molecular weights of viroids is of great importance in elucidating their structure. This determination is difficult, because the agents occur in infected tissue in very small amounts and are, therefore, difficult to separate from host RNA and to purify in amounts sufficient for conventional biophysical analyses.

A concept elaborated by Loening (1967) made it feasible to determine the molecular weight of PSTV, using infectivity as the sole parameter. Combined sedimentation and gel electrophoretic analyses conclusively showed that the infectious RNA has a very low molecular weight (Diener, 1971b). A value of 5 x 10^4 daltons was compatible with the experimental results (Diener, 1971b). This conclusion was confirmed by the ability of PSTV to penetrate into polyacrylamide gels of high concentration (small pore size), from which high molecular weight RNA molecules are excluded (Diener and Smith, 1971).

On the basis of its electrophoretic mobility in 5% polyacrylamide gels, Semancik and Weathers (1972) came to the conclusion that the agent causing citrus exocortis disease is also a low-molecular-weight RNA. They estimated that the RNA has a molecular weight of 1.25 x 10^5 Daltons. Sanger (1972), on the other hand, estimated that the exocortis disease agent has a molecular weight of 5 to 6 x 10^4

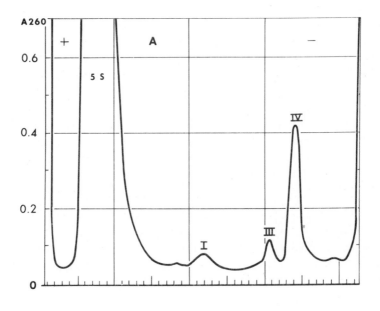

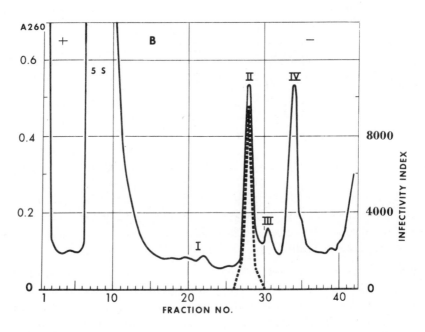

daltons.

Diener and Lawson (1973) showed by a combination of iso-kinetic density-gradient centrifugation and electrophoresis in 20% poly-acrylamide gels that the agent of chrysanthemum stunt disease is a low-molecular-weight RNA similar to, but distinct from, PSTV.

PURIFICATION OF PSTV.

In all experiments so far described, PSTV was identified solely by its biological activity; and no clearly and consistently recognizable ultraviolet light-absorbing component was correlated with infectivity distribution in sucrose gradients or in polyacrylamide gels. Detailed characterization of PSTV requires its isolation in amounts sufficient for conventional biophysical and biochemical analyses. Large-scale isolation of PSTV, together with improvements in separation techniques, have made this goal attainable (Diener, 1972a).

As shown in Figure 1A, electrophoresis in 20% polyacrylamide gels of highly concentrated, low-molecular-weight RNA preparations from healthy plants discloses, aside from 5 S RNA, at least three RNA species of low molecular weight (I, III, IV). Electrophoresis of identically prepared samples from PSTV-infected plants reveal the same components; namely 5 S RNA and components I, III, and IV; but, in addition, another prominent ultraviolet light-absorbing component, II, is evident (Fig 1B). Bioassays of individual gel slices demonstrated that infectivity coincides with component II (Fig 1B). This coincidence, the high level of infectivity, and the fact that component II is not recognizable in preparations from healthy leaves constitutes strong evidence that component II is PSTV.

Fig. 1. A. Ultraviolet light-absorption profile of RNA preparation from healthy tomato leaves after electrophoresis in a 20% polyacrylamide gel for 7.5 hr at 4°C (5mA per tube, constant current).

Fig. 1. B. Ultraviolet light-absorption (———) and infectivity distribution (·····) of RNA preparation from PSTV-infected tomato leaves after electrophoresis in a 20% polyacrylamide gel (same conditions as in A). 5 S = 5 S ribosomal RNA; I, III, IV = unidentified minor components of cellular RNA; II = PSTV; A_{260} = Absorbance at 260 nm. Electrophoretic movement from right to left. (From Diener, 1972a).

Pure PSTV has now been prepared using electrophoresis in 20% polyacrylamide gels as the last step in the purification scheme, followed
by elution from gel slices and reconcentration of the RNA by chromatography on hydroxyapatite (Diener, 1973). Quantities so far produced
were sufficient to determine several properties of the RNA by conventional means.

THERMAL DENATURATION PROPERTIES.

The total hyperchromicity shift of PSTV in 0.01 x SSC (0.15 M
sodium chloride-0.015 M citrate, pH 7.0) is about 24% and the T_M
about 50°C (Diener, 1972a). The thermal denaturation curve indicates
that PSTV is not a regularly base-paired structure, such as double-
stranded RNA, since in this case, denaturation would be expected to
occur over a narrower temperature range and at higher temperatures
(Miura *et al.*, 1966). PSTV could, however, be an irregularly base-
paired, single-stranded RNA molecule, similar to transfer RNA, in
which single-stranded regions alternate with base-paired regions.

MOLECULAR WEIGHT OF PSTV.

With the availability of purified PSTV, a redetermination of its
molecular weight based, not on its biological activity, but on its
physical properties, became possible. For this purpose, a method
described by Boedtker (1971) appeared particularly promising, as it
permits the determination of the molecular weight of an RNA independent of its conformation.

Application of this method to PSTV led to a molecular weight estimate of 7.5-8.5 x 10^4 Daltons (Diener and Smith, 1973).

VISUALIZATION OF PSTV.

In view of the purity of available PSTV preparations, it appeared
feasible to visualize PSTV by electron microscopy and to determine its
molecular weight by direct length measurements of the RNA in electron
micrographs. This, indeed, proved feasible (Sogo *et al.*, 1973).

Figure 2 shows an electron micrograph of a mixture of a double-
stranded DNA; namely, coliphage T_7-DNA, and PSTV. Length measurements indicated that T_7-DNA is about 280 times longer than PSTV. It
is also apparent that the width of PSTV is similar to that of T_7-DNA.
Assuming a molecular weight of T_7-DNA of 25 x 10^6 Daltons, and assuming that PSTV in urea is formed by two more or less base-paired strands (either as a hairpin or a double helix), then one obtains a molecular weight estimate for PSTV of 8.9 x 10^4 Daltons.

In other experiments, mixtures of PSTV and a single-stranded viral

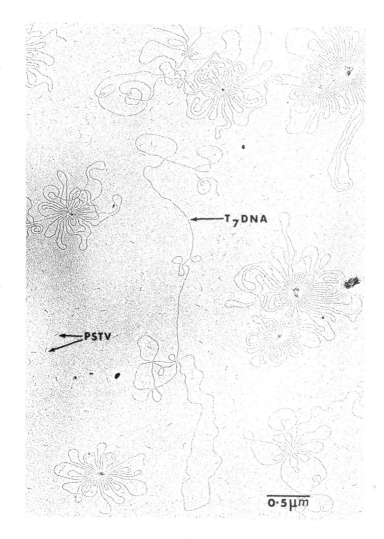

Fig. 2. Electron micrographs of PSTV mixed with a double-stranded DNA, coliphage T_7-DNA. Native T_7-DNA (0.8µg/ml) was mixed with purified PSTV (0.4µg/ml) previously heated for 10 min at 63°C in the presence of 8 M urea, followed by quenching in ice water. Note that double-stranded T_7-DNA and PSTV have similar widths. (Courtesy: T. Koller and J.M. Sogo, Swiss Federal Institute of Technology, Zurich).

RNA were examined and measured. PSTV appeared thicker than this single-stranded viral RNA, and from length comparisons, a molecular weight of 7.9×10^4 Daltons was obtained (Sogo *et al.*, 1973).

The molecular weight estimates obtained by electron microscopy are, therefore, in excellent agreement with the values obtained from analysis of heat-denatured, formylated PSTV in polyacrylamide gels.

INACTIVATION BY ULTRAVIOLET LIGHT.

In view of the low molecular weight of PSTV, it was of interest to determine its sensitivity to irradiation with ultraviolet light. Although one might expect that a small molecule, such as PSTV, would be considerably more resistant to ultraviolet irradiation than a conventionally sized viral RNA or DNA, the effect of size on ultraviolet sensitivity of nucleic acids is not well understood (Adams, 1970).

Exposure of purified PSTV, of tobacco ringspot virus (TRSV), and of its satellite (SAT) to ultraviolet radiation of 254 nm showed that the inactivation dose for PSTV and SAT is 70 to 90 times as large as that for TRSV (Diener *et al.*, 1974). Although other explanations are possible, this marked difference in sensitivity to ultraviolet radiation is most likely a consequence of the smaller size (smaller target volume) of PSTV and SAT-RNA, as compared with TRSV-RNA.

STRUCTURE OF PSTV.

Although the low molecular weight of PSTV has been conclusively demonstrated, in light of present knowledge a decision as to its exact structure cannot be made unambiguously. In some analytical systems, PSTV displays properties typical of double-stranded RNA, in others of single-stranded RNA.

Two models are compatible with known properties of PSTV: a) The RNA may be a single-stranded molecule with some sort of hairpin structure, involving extensive base-pairing, or b) the RNA may be a double-stranded, but incompletely base-paired molecule.

CONCLUSIONS

With the evidence now at hand, little doubt is left that the potato spindle tuber disease is caused by a low-molecular-weight RNA which, when introduced into susceptible host cells, is able to replicate and to produce the characteristic syndrome of the disease. In view of the small amount of genetic information that PSTV introduces into its host cells, one might plausibly assume that PSTV is analogous to a satellite RNA that requires a helper virus for its own replication. However, efforts to demonstrate the presence of such a helper virus in uninoculated tomato plants gave negative results (Diener, 1971b). In view of these and other results (Diener, 1971b;

Diener *et al.*, 1972), it appears most unlikely that a conventional helper virus is necessary for the replication of PSTV.

REPLICATION OF VIROIDS

By what mechanism an RNA of such a small size is replicated in susceptible host cells is, at present, unknown but, as discussed previously (Diener, 1971b; 1972b, c), two schemes are most readily compatible with present views on cellular nucleic acid synthesis. Either scheme may derive support from seemingly unrelated recent discoveries.

The first scheme postulates that PSTV is replicated on a DNA template. Since such a template is not likely to be present in uninoculated plants, it would have to be synthesized as a consequence of infection with PSTV. Implicit in this scheme, therefore, is the assumption that RNA-dependent DNA polymerases occur in normal cells. Scolnick *et al.* (1971) indeed claimed that such enzymes occur in normal cells, but this claim is a matter of controversy (Schlom *et al.*, 1971).

Coffin and Temin (1971), on the other hand, demonstrated ribonuclease-sensitive DNA polymerase activity in preparations from uninfected rat cells and showed that the product of endogenous DNA polymerase is DNA. The authors suggested that this enzymic activity in uninfected cells may be due to a chance association between a cell DNA polymerase and cell RNA, or may be related to an RNA-directed DNA polymerase involved in normal development.

The second scheme presumes that PSTV is replicated independently of DNA; that is to say, that its replication is analogous to that of certain viral RNA's. Implicit in this scheme is the assumption that RNA-dependent RNA polymerases occur in normal cells.

For a number of years, it has been known that certain bacterial RNA polymerases synthesize RNA with both RNA and DNA primers (Fox *et al.*, 1964).

Evidently, if similar enzymes occur in higher organisms and accept a similarly wide range of RNA's as primers, such enzymes may be responsible for the replication of viroids. Two reports (Astier-Manifacier and Cornuet, 1971; Duda *et al.*, 1973) suggest that RNA-directed RNA polymerases may, indeed, occur in apparently healthy plants.

Implicit in this scheme also is the expectation that RNA structures analogous to viral replicating forms; i.e., double-stranded RNA's, should occur in normal cells. A number of investigators have, indeed, discovered such structures in a variety of apparently uninfected cells (Montagnier, 1968a; Duesberg and Colby, 1969; Stern and Friedman, 1970; Stollar and Stollar, 1970; and others), as well as in tomato leaves (Lewandowski et al., 1971). Although other interpretations are possible, it has been suggested that "some self-replicating RNA's may have become part of cell genetic information" (Montagnier, 1968b).

IMPLICATIONS

The evidence now available leaves little doubt that viroids are pathogens which drastically differ from any other pathogenic agents, including viruses. They are the smallest known agents of infectious disease.

Although viroids so far identified cause diseases of higher plants, similar agents may exist in other forms of life; and it appears reasonable to search for viroids in the many instances where viral aetiology of an infectious disease has been assumed, but where no causative agent has ever been isolated and characterized.

One case in point is the agent of the scrapie disease of sheep (Diener, 1972d), but there are, undoubtedly, many more.

Evidently, only future work will determine whether some animal diseases of obscure aetiology are caused by agents resembling presently known viroids affecting higher plants.

REFERENCES

Adams, D.H. (1970). The nature of the scrapie agent. A review of recent progress. *Pathologie et Biologie*. 18, 559-577.

Astier-Manifacier, S., and Cornuet, P. (1971). RNA-dependent RNA polymerase in Chinese cabbage. *Biochimica et Biophysica Acta*. 232, 484-493.

Boedtker, H. (1971). Conformation independent molecular weight determination of RNA by gel electrophoresis. *Biochimica et Biophysica Acta*. 240, 448-453.

Coffin, J.M., and Temin, H.M. (1971). Ribonuclease-sensitive deoxyribonucleic acid polymerase activity in uninfected rat cells and rat cells infected with Rous sarcoma virus. *Journal of Virology*. 8, 630-642.

Diener, T.O. (1971a). Potato spindle tuber virus. III. Subcellular location of PSTV-RNA and the question of whether virions exist in extracts or *in situ*. *Virology*. 43, 75-89.

Diener, T.O. (1971b). Potato spindle tuber "virus". IV. A replicating, low molecular weight RNA. *Virology*. 45, 411-428.

Diener, T.O. (1971c). A plant virus with properties of a free ribonucleic acid: Potato spindle tuber virus. In *Comparative Virology*. K. Maramorosch and E. Kurstak (eds), p. 433-478. Academic Press, New York.

Diener, T.O. (1972a). Potato spindle tuber viroid. VIII. Correlation of infectivity with a ultraviolet light-absorbing component and thermal denaturation properties of the RNA. *Virology*. 50, 606-609.

Diener, T.O. (1972b). Potato spindle tuber viroid: A novel type of pathogen. *Perspectives in Virology*. 8, 7-30.

Diener, T.O. (1972c). Viroids. *Advances in Virus Research*. 17, 295-313.

Diener, T.O. (1972d). Is the scrapie agent a viroid? *Nature New Biology*. 235, 218-219.

Diener, T.O. (1973). A method for the purification and reconcentration of nucleic acids eluted or extracted from polyacrylamide gels. *Analytical Biochemistry*. 55, 317-320.

Diener, T.O., and Lawson, R.H. (1973). Chrysanthemum stunt: A viroid disease. *Virology*. 51, 94-101.

Diener, T.O., and Raymer, W.B. (1967). Potato spindle tuber virus: A plant virus with properties of a free nucleic acid. *Science*. 158, 378-381.

Diener, T.O., and Raymer, W.B. (1969). Potato spindle tuber virus: A plant virus with properties of a free nucleic acid. II. Characterization and partial purification. *Virology*. 37, 351-366.

Diener, T.O., and Raymer, W.B. (1971). Potato spindle tuber "virus". *Descriptions of Plant Viruses*. No. 66. Kew, Surrey, England: Commonwealth Mycological Institute and Association of Applied Biologists. 4pp.

Diener, T.O., and Smith, D.R. (1971). Potato spindle tuber viroid. VI. Monodisperse distribution after electrophoresis in 20 percent polyacrylamide gels. *Virology*. 46, 498-499.

Diener, T.O., and Smith, D.R. (1973). Potato spindle tuber viroid. IX. Molecular weight determination by gel electrophoresis of formylated RNA. *Virology*. 53, 359-365.

Diener, T.O., Smith, D.R., and O'Brien, M.J. (1972). Potato spindle tuber viroid. VII. Susceptibility of several solanaceous plant

species to infection with low molecular weight RNA. *Virology.*
48, 844-846.

Diener, T.O., Schneider, I.R., and Smith, D.R. (1974). Potato spindle
tuber viroid. XI. A comparison of the ultraviolet light sensiti-
vities of PSTV, tobacco ringspot virus, and its satellite.
Virology. 57, 577-581.

Duda, C.T., Zaitlin, M., and Siegel, A. (1973). *In vitro* synthesis
of double-stranded RNA by an enzyme system isolated from tobacco
leaves. *Biochimica et Biophysica Acta.* 319, 62-71.

Duesberg, P.H., and Colby, C. (1969). On the biosynthesis and
structure of double-stranded RNA in vaccinia virus-infected cells.
Proceedings of the National Academy of Sciences, U.S. 64, 396-403.

Fox, C.F., Robinson, W.S., Haselkorn, R., and Weiss, S.B. (1964).
Enzymatic synthesis of ribonucleic acid. III. The ribonucleic
acid-primed synthesis of ribonucleic acid with *Micrococcus
lysodeikticus* ribonucleic acid polymerase. *Journal of Biological
Chemistry.* 239, 186-195.

Lawson, R.H. (1968). Some properties of chrysanthemum stunt virus.
Phytopathology. 58, 885.

Lewandowski, L.J., Kimball, P.C., and Knight, C.A. (1971). Separation
of the infectious ribonucleic acid of potato spindle tuber virus
from double-stranded ribonucleic acid of plant tissue extracts.
Journal of Virology. 8, 809-812.

Loening, U.E. (1967). The fractionation of high-molecular-weight
ribonucleic acid by polyacrylamide-gel electrophoresis. *Biochemical
Journal.* 102, 251-257.

Miura, K.I., Kimura, I., and Suzuki, N. (1966). Double-stranded
ribonucleic acid from rice dwarf virus. *Virology.* 28, 571-579.

Montagnier, L. (1968a). Présence d'un acide ribonucléique en double
chaîne dans des cellules animales. *Compte rendus des Séances de
l'Académie des Sciences. Serie D.* 267, 1417-1420.

Montagnier, L. (1968b). Replication of viral RNA. *Symposia of the
Society for General Microbiology.* 18, 125-148.

Sänger, H.L. (1972). An infectious and replicating RNA of low molecular weight: The agent of the exocortis disease of citrus.
Advances in Biosciences. 8, 103-116.

Schlom, J., Spiegelman, S., and Moore, D. (1971). RNA-dependent DNA
polymerase activity in virus-like particles isolated from human
milk. *Nature* (London). 231, 97-100.

Scolnick, E.M., Aaronson, S.A., Todaro, G.J., and Parks, W.P. (1971).

RNA dependent DNA polymerase activity in mammalian cells. *Nature* (London). 229, 318-321.

Semancik, J.S., and Weathers, L.G. (1968). Exocortis virus of citrus: Association of infectivity with nucleic acid preparations. *Virology.* 36, 326-328.

Semancik, J.S., and Weathers, L.G. (1972). Exocortis disease: Evidence for a new species of "infectious" low molecular weight RNA in plants. *Nature New Biology.* 237, 242-244.

Singh, R.P., and Bagnall, R.H. (1968). Infectious nucleic acid from host tissues infected with potato spindle tuber virus. *Phytopathology.* 58, 696-699.

Sogo, J.M., Koller, T., and Diener, T.O. (1973). Potato spindle tuber viroid. X. Visualization and size determination by electron microscopy. *Virology.* 55, 70-80.

Stern, R., and Friedman, R.M. (1970). Double-stranded RNA synthesized in animal cells in the presence of actinomycin D. *Nature* (London). 226, 612-616.

Stollar, V., and Stollar, B.D. (1970). Immunochemical measurement of double-stranded RNA of uninfected and arbovirus-infected mammalian cells. *Proceedings of the National Academy of Sciences, U.S.* 65, 993-1000.

Zaitlin, M., and Hariharasubramanian, V. (1972). Gel electrophoretic analysis of proteins from plants infected with tobacco mosaic and potato spindle tuber viruses. *Virology.* 47, 296-305.

N.B. Figure 1 is reproduced by permission from:
    Diener, T.O., *Virology.* 50, p. 607, Fig. 1 (1972).

RADIOACTIVE LABELLING OF VIROID-RNA

H.L. SÄNGER AND K. RAMM

Arbeitsgruppe Pflanzenvirologie,
Justus Liebig-Universität,
D 63 Giessen, Schubertstrasse 1.

The synthesis of the free infectious RNA of the potato
spindle tuber viroid (PSTV) and of the viroid of the
exocortis disease of citrus (ExCV) in their specific host
plants is unusually temperature-dependent. At growth temperatures below $24°C$ viroid-RNA accumulates to concentrations which are less than 0.001% of the total RNA extracted
from plant tissue. The concentration may rise to about
0.3% of the total extractable RNA if the host plants are
grown at temperatures between $30-35°C$ for viroid replication.
At this concentration viroid-RNA can be easily detected in
cylindrical 5% polyacrylamide gels by spectrophotometric
scanning at 260 nm. The incorporation in infected plants
of ^{32}P into viroid-RNA under optimal conditions allowed the
detection of the viroid band in preparative slab gels by
autoradiography, which greatly facilitates the purification
of viroid-RNA. Viroid-RNA could also be demonstrated in
individual leaves by the incorporation of ^{3}H-uridine as
compared with the incorporation of ^{14}C-uridine in healthy
control leaves. "Long-term" incorporation studies suggest
that viroid-RNA synthesis has no dramatic effect on the RNA
metabolism of the host cell.

Progress in the study on the structure and function of viroids is
slow and mainly limited by the extremely low concentrations of viroid-
RNA present in viroid-infected plant cells (Diener and Raymer, 1969;
Raymer and Diener, 1969; Diener, 1971; Sänger, 1972; Semancik and
Weathers, 1972). Accordingly, large quantities of total RNA from
infected plants have to be processed in order to concentrate the
minute fraction of viroid-RNA to microgram amounts (Diener, 1972).
It has been estimated that it comprises only around 0.02% of the
total nucleic acids extracted directly from plant tissue (Semancik
et al., 1973).

Considering that the availability of radioactively labelled viroid-
RNA would facilitate investigations of both viroid-structure and

viroid-replication we have studied the conditions under which radio-active viroid-RNA can be obtained.

MATERIAL AND METHODS

VIROID CULTURE AND BIOASSAY.

The viroid of the exocortis disease of citrus (ExCV) was cultured and bioassayed in *Gynura aurantiaca* DC, whereas Rutgers tomato plants were used for the culture and bioassay of the viroid of the potato spindle tuber disease (PSTV). The stock cultures of both viriods were kindly supplied by Dr. Weathers, Riverside and Dr. Diener, Beltsville, respectively. The bioassay for ExCV was carried out by inoculating the plant stem with a standardized needle-puncture method as previously described (Sänger, 1972). Rutgers tomato plants were used in the four-leaf stage for the bioassay of PSTV and inoculated by leaf rubbing with the aid of carborundum. In both systems ten plants and two or three dilutions were used for each sample to be tested.

The influence of temperature on viroid infection under greenhouse conditions was studied in temperature-controlled growth chambers with a constant illumination of about 6000 Lux for 16 hrs per day using fluorescent lamps especially designed for plant illumination (Fluora lamps by Osram). According to the distribution of their spectral irradiance this illumination corresponds to about 36 000 Lux daylight-equivalents. The plants were raised at 20-22°C, adapted to the corre-sponding environmental condition for four days, inoculated and kept at their temperature until assayed.

EXTRACTION OF RNA.

The leaf material to be used was shock-frozen with liquid nitrogen, pulverized in the frozen state and stored for at least two weeks at -30°C. Total nucleic acids were extracted with the aid of phenol and sodium dodecylsulphate and concentrated by precipitation with ethanol and with cetyltrimethylammonium bromide as previously described (Sänger, 1972). In addition, the clarified solution of dissolved nucleic acids was subjected to salt fractionation by adding 2 volumes of 4 M LiCl to the final preparation and letting it stand overnight for precipitation. The precipitate was collected by low speed centri-fugation and contained the "2 M LiCl-insoluble RNA" which is predomi-nantly ribosomal RNA. To the resulting supernatant 2.5 volumes of ethanol were added in order to precipitate the "2 M LiCl-soluble RNA" which contains transfer RNA, 5S RNA, several nuclear RNA species, DNA, and also the viroid RNA. The final precipitates were dissolved in

0·01 M TES buffer pH 7.5 and stored frozen until used.

LABELLING WITH ^{32}P

The host plants to be used in the labelling experiments were grown
in the greenhouse at 20-22°C in 8 cm plastic pots in a sandy soil mix.
For the uptake of ^{32}P their root system was washed free from adhering
soil under running tap water. Then each plant was put with its roots
into siliconized scintillation vials containing 20 ml of distilled
water to which 200-400 µl of a stock solution of ^{32}P as orthophosphate
(PBS. 1. Radiochemical Centre Amersham) had been added. During the
^{32}P-experiment the plants were kept with their vials sitting in plastic racks inside a desk-size growth chamber made out of "Plexiglas"
(10 mm thick) with an outer lead shielding (3 mm thick). This provided practically full protection from the heavy radiation emitted from
plants and vials. The plants were illuminated with 20 000 Lux daylight-equivalents for 16 hrs per day and temperature was held constant
to ± 1°C at a relative humidity of 90 percent. During the experiment
the plants took up 5 ml of solution in average per day; accordingly,
the vials were refilled twice daily with a dilute nutrient solution.

LABELLING WITH ^{3}H- AND ^{14}C-URIDINE.

Labelling with ^{3}H- and ^{14}C-uridine (TRK 178 and CTA 315, Radiochemical Center, Amersham) was carried out with individual leaves and
the RNA precursor was allowed to be taken up in 100-200 µl via the
petiole out of small conical test tubes. After the uptake of two
subsequent washings with 200 µl water the tubes were filled up to a
content of 2 ml water in which the leaves were kept under identical
conditions as described for the ^{32}P-labelling.

GEL-ELECTROPHORESIS.

Electrophoresis was performed in vertical polyacrylamide gel slabs
(23 x 26 cm, 5 mm thick) in tris-borate-EDTA-buffer pH 8.3 (Peacock
and Dingman, 1967) with 0.2% sodium dodecylsulphate. Gels were prerun
for 30 mins at 50 mA/gel with a mixture of bromophenol-blue and pyro-
nin G as marker dyes. The samples were dissolved in the electrode
buffer containing 15% sucrose and applied to the slots in the gel slab
in a volume of 0.2 - 0.5 ml. Electrophoresis was carried out in a
coldroom at +7°C for 22-26 hrs at 50 mA/gel at which time the marker
dyes had migrated about 20 and 25 cm, respectively. Electrophoresis
in cylindrical gels (15 cm long, 6 mm in diameter) was performed under
essentially the same conditions as the preparative slab gel runs.
These gels were scanned at 260 nm with a Gilford spectrophotometer
equipped with a 20 cm gel-scanner and sliced in 1 mm disks on a

Mickle model 140 gel slicer.

AUTORADIOGRAPHY AND COUNTING.

After electrophoresis the gel slabs were autoradiographed for the localization of the RNA species. One of the two glass plates holding the gel in position was carefully removed; then the gel was covered with thin polyethylene household wrap, sealed and fastened with plastic tape and marked with radioactive ink. X-ray film (Agfa-Gevaert, type Curix RP1, 24 x 30 cm) was put on top of the sealed gel in the dark room, covered with a glass plate, clamped together and wrapped light-tight in aluminium foil. This pack was kept in the refrigerator for exposure, the time of which was usually between 2 and 24 hrs according to the radioactivity present and the intensity of the blackening wanted.

The radioactivity of ^{32}P-containing preparations was counted in aqueous solution as Cerenkov-counts (Haviland and Bieber, 1970) at a 2 σ error of 0.2%. ^{3}H- and ^{14}C-containing gel slices were solubilized with Soluene 350 (Packard) for 2 hrs at +60°C and dissolved and counted in a toluene scintillation mixture containing Scintimix 1 (91% PPO and 9% bis-MSB, Koch-Light Laboratories Ltd.) as scintillator at a 2 σ error of 2%.

EXTRACTION OF RNA *FROM THE GELS.*

For infectivity tests RNA was eluted overnight from gel-slices at 4°C in 0.1 M TES buffer.

For preparative purposes gel bands were cut out from the slab gels according to the autoradiograph. The gel pieces were put into scintillation vials and radioactivity was counted directly in the gel as Cerenkov counts. Then, the gel was thoroughly homogenized in 0.1 M TS buffer (0.1 M Tris, 0.1 M NaCl) pH 7.5. The homogenate was centrifuged at 5000 x g and the supernatant was carefully pipetted off. This extraction procedure was repeated twice more. The dilute radioactive RNA present in the three supernatants was adsorbed to a hydroxyapatite column, eluted with 0.25 M phosphate buffer, precipitated with cetyltrimethylammonium bromide, converted back to the sodium salt and dissolved in 0.01 M TES buffer for further analysis.

RESULTS

VIROID-INFECTION AND ENVIRONMENTAL CONDITIONS.

In earlier experiments no UV-absorbing component could be found to coincide with the distribution of infectivity in polyacrylamide gels in the case of both PSTV (Diener, 1971; Diener and Smith, 1971) and

ExCV (Sänger, 1972). After the extraction of RNA from large quantities of viroid-infected tissue highly concentrated preparations were
obtained in which a specific UV-absorbing peak resembling viroid-RNA
could be demonstrated for PSTV (Diener, 1972) and for ExCV (Semancik
and Weathers, 1972; Semancik et al., 1973). Until Autumn 1973 we
failed to obtain this viroid-specific UV-absorbing peak despite repeated attempts using concentrated RNA-preparations extracted from two
to four kilograms of infected and symptom-bearing plant tissue. Yet,
the bioassay of the gels on which these RNA preparations were analysed showed the presence of the infectious viroid-RNA at the expected
position and the eluates from the corresponding gel slices had dilution end points of 10^{-3} to 10^{-5}. Since the gel pattern of the plant
RNA appeared to be normal our failure could not be related to an inadequate extraction and concentration of the RNA. Therefore, we presumed that unfavourable environmental conditions during the propagation of the viroids in their hosts were the cause of the extremely
low concentration of viroid-RNA in the corresponding preparations.

This presumption was supported by the observation that the success
of viroid infection and the expression of symptoms was greatly en-
hanced in ExCV-infected Gynura plants as well as in PSTV-infected
tomato plants during longer periods of "hot" summer weather which
caused the greenhouse temperature to rise above $30^{\circ}C$ for 8 to 10 hrs
per day. In addition, the incubation period was reduced from several
weeks at temperatures around $20^{\circ}C$ to about 1 week at temperatures
above $30^{\circ}C$ in summer. The result of a systematic study of this relationship is shown in Table 1.

It clearly substantiates that at elevated greenhouse temperatures
the success of viroid infection is increased while the period of incubation is drastically reduced. Parallel to this the development of
the symptoms and their ultimate severity are also positively correlated with the temperature at which the host plant is growing, although this relationship is less striking.

Furthermore, illumination has a pronounced effect on the period of
incubation and on the severity of the disease symptoms. Although no
systematic study was made, it was found that, at a constant greenhouse temperature of $20-22^{\circ}C$ and at an illumination with 36 000 Lux
day-light equivalents (DLE), the mean incubation period of ExCV in
Gynura is about 30 days, as judged by the first appearance of symptoms. If the illumination is reduced to 15 000 Lux DLE the incubation
period is extended to about 42 days and at 1000 Lux DLE it is increa-

Table 1

The influence of growth-temperature on the success of
infection and on the length of the incubation period of ExCV in
Gynura aurantiaca and of PSTV in Rutgers tomato

Growth temperature in °C		ExCV in *Gynura*		PSTV in tomato	
Day	Night	Infectivity[1]	Mean incubation-period in days	Infectivity[1]	Mean incubation-period in days
15	15	1/10	65	3/10	45
20	18	2/10	35	5/10	28
24	20	6/10	23	8/10	17
28	23	8/10	14	10/10	12
35	28	10/10	10	10/10	8

[1]Number of infected plants over number of inoculated plants.

sed to about 60 days. Parallel to this the severity of the disease
symptoms is significantly reduced. This effect of illumination can
also be observed with PSTV in tomato, although it is less pronounced
in this case.

From all these observations we deduced that growth temperature and
illumination were the limiting factors for viroid propagation under
our normal greenhouse conditions. In fact our greenhouse temperatures
range between 18° and 22°C most of the year combined with low (500 -
5000 Lux) to medium (6000 - 15 000 Lux) light intensities. On the
other hand, greenhouse temperatures above 25°C are usually associated
with high light intensities (above 35 000 Lux) from sunshine. Consequently we now provide additional light and heating in our greenhouse throughout the year except during bright sunshine.

ANALYSIS ON POLYACRYLAMIDE GELS.

Direct evidence for the pronounced effect of temperature on viroid-
synthesis was obtained when the "2 M LiCl-soluble RNA" from viroid-
infected plants grown at different temperatures was run on 5 % polyacrylamide gels. On such gels viroid-RNA, if present at a sufficient
concentration, appears as a new and distinct UV-absorbing peak, absent
in healthy tissue preparations. Evidence for the viroid nature of
this novel RNA comes from the bioassay of gel slices. Experiments
showed that the infectivity maximum coincides with the maximum of the
UV-absorbance of this peak. The viroid nature of the peak is confirmed by comparing infectivity tests of fractionated slab gels with the
autoradiograms of these gels (see Fig 4).

Since the viroid-RNA peak can be clearly distinguished from other
cellular RNA species, including a neighbouring 8S-RNA, by gel scanning at 260 nm, the analysis on a 5% gel provides a simple analytical
technique for studying the effect of growth temperatures on viroid-
RNA synthesis.

The results of a systematic study on the influence of temperature
on the synthesis of PSTV-RNA in Rutgers tomatoes is shown in Figure 1.
It should be noted, that 25 infected tomato plants were grown for
4 weeks at the given temperature and that for each temperature the
equivalent of the "2 M LiCl-soluble RNA" from a sample of 50 g of
total leaf tissue was dissolved in 100 µl of 0.01 M TES-buffer pH
7.5 and analysed on one gel. Therefore, the PSTV-RNA extracted together with the cellular RNA species is actually the RNA that has
been synthesised and accumulated during this period of time. It
becomes evident, that under the special conditions of the analysis

236

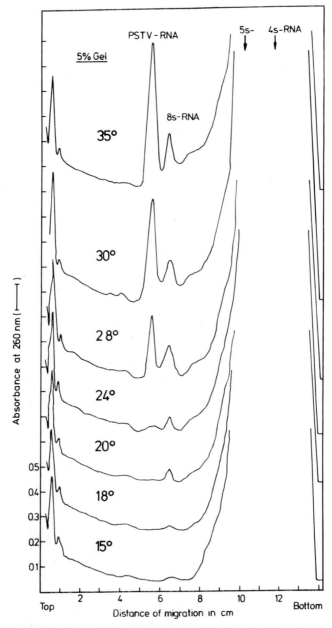

-Fig. 1. Electrophoretic analysis on 5% polyacrylamide gels of "2 M LiCl-soluble RNA" from PSTV-infected tomatoes grown for four weeks at different temperatures. On each gel the RNA-equivalent from 50 g leaf tissue was run. The increase of the concentration of PSTV-RNA with increasing growth temperature is evident.

applied PSTV-RNA is accumulating in Rutgers tomato plants to concentrations discernible in the gel as UV-absorbing peaks only at temperatures above $24^{\circ}C$. On the other hand, the unusually high concentration of PSTV-RNA synthesised at temperatures above $30^{\circ}C$ allows the detection of the viroid peak from the equivalent of the "2 M LiCl-soluble RNA" from 1 to 2 g of leaf tissue only. In this case, however, the gels need to be scanned at high sensitivity with 0.1-0.5 OD_{260} units at full scale.

The bioassay of gel slices showed that the infectivity distribution in the gel coincided with the UV-absorbing peak resembling the PSTV-RNA. It is interesting to note that in addition to the increased synthesis of viroid-RNA there is also an apparent increase in the concentration of the 8S-plant RNA, which can be extracted from healthy as well as from viroid-infected plant tissue. The nature and the function of this plant RNA is not known and at present it can not be clearly decided whether its increase with temperature is based on a stimulation of its synthesis or on the release from, or the breakdown of, a possible functional complex.

Similar results were obtained when ExCV was propagated in *Gynura aurantiaca* under identical conditions. It was found, however, that the yield of ExCV-RNA as based on the fresh weight of leaf tissue was in average less than 50% as compared to the yield of PSTV-RNA from Rutgers tomato.

RADIOACTIVE LABELLING WITH ^{32}P *OF* PSTV-RNA.

The finding that temperature may influence viroid-synthesis significantly led us to investigate this relationship by radioactive labelling with ^{32}P at $18-20^{\circ}C$ and $30-32^{\circ}C$. We expected that at these two extremes of temperature differences should be clearly discernible.

In one labelling experiment we propagated PSTV in tomato plants at $18-20^{\circ}C$ and 10 000 Lux daylight-equivalents. The infected plants were used 20 days after inoculation at which time they showed slight symptoms of systemic infection. Ten uninfected plants were used as a control. To each plant 1.0 mCi ^{32}P was given as orthophosphate under the conditions described in Materials and Methods. After 5 days of exposure to the ^{32}P the plants were harvested, shock-frozen and the RNA was phenol-extracted, LiCl-fractionated and analysed on 5% and 10% gels. Figure 2 shows the result of the analysis on a 5% gel. It can be seen from the autoradiograph that there is no difference between the RNA pattern from healthy (H) and from PSTV-infected (I) plants. As expected, the supernatant from the 2 M LiCl fraction-

238

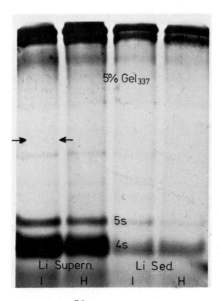

Fig. 2. Autoradiograph of ^{32}P-labelled and LiCl-fractionated nucleic
   acid preparations from PSTV-infected (I) and healthy (H) tomato
   plants after electrophoresis on a 5% polyacrylamide gel.

   The labelling was carried out for 5 days at $18-20^\circ$C and the RNA-
   equivalent of 5 g leaf tissue was applied to each slot.

   The arrows indicate the position of the infectious PSTV-RNA as
   determined by the bioassay of gel slices. For further details
   see text.

ation contains predominantly transfer RNA, 5S-RNA and DNA as the
major nucleic acid species soluble in 2 M LiCl. The DNA barely pene-
trates into the gel and causes the intense blackening at the origin.
One would expect that the PSTV-RNA newly synthesised during the experiment becomes ^{32}P-labelled and if so it should actually appear as
a distinct band halfway in between the start and the 5S-RNA, just
above the 8S-plant RNA. Even prolonged exposure of the X-ray film
did not reveal any additional band in the preparation from viroid-
infected plant tissue.

The bioassay carried out with eluates from slices of the gel in
Figure 2 showed, however, that a minor amount of unlabelled PSTV-RNA
(dilution end point 10^{-3}) was present in the part of the gel (see
arrow) in which the "2 M LiCl-soluble RNA" (= Li-Supern.) from PSTV-
infected plants had been run. No infectivity was detectable in any
gel slices from the gel parts in which the three other preparations
(Li supern. H, Li sed. H, Li sed I) had been electrophoresed. It
should be noted that the "2 M LiCl-insoluble RNA" (= Li-Sed.) which

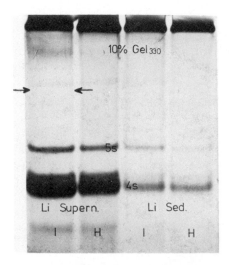

Fig. 3. Autoradiograph of ^{32}P-labelled and LiCl-fractionated nucleic acid preparations from PSTV-infected (I) and healthy (H) tomato plants after electrophoresis on a 10% polyacrylamide gel.

The labelling was carried out for 5 days at 18-20°C and the RNA-equivalent of 5 g leaf tissue was applied to each slot.

For further details and for comparison see Figure 2 in which the analysis of samples on a 5% gel is shown.

is expected to consist mostly of ribosomal (r) RNA still contains small amounts of 5S-RNA, t-RNA and the other minor RNA species. The ribosomal RNA itself, however, migrates only a few millimeters into the top of the gel. The analysis of portions of same RNA preparations on a 10% gel is shown in Figure 3. Again, there is no difference in the pattern of RNA detectable and prolonged time of exposure did not reveal any further details of importance. In this gel the maximum viroid infectivity (see arrow) was detected close to the region of the 8S-plant RNA which is clearly visible in the upper third of the gel. These results show, that practically no viroid-RNA is synthesised at temperatures between 18 and 20°C.

For comparison the influence of a temperature-range favourable to viroid synthesis was tested. Tomato plants from the same batch as used in the previous experiment were grown at 22°C and 10 000 Lux daylight-equivalents, adapted to 30-32°C two days before labelling and the ^{32}P-labelling was carried out from the 20th to the 26th day after inoculation at 30-32°C. The analysis on a 5% gel of the nucleic acids synthesised at "high temperature" in healthy and in PSTV-infected tomatoes is shown in Figure 4. It can be seen that under these

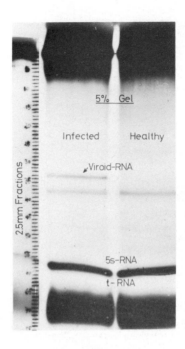

Fig. 4. Autoradiograph of ^{32}P-labelled "2 M LiCl-soluble RNA" from PSTV-infected and healthy tomato plants after electrophoresis on a 5% polyacrylamide gel.

The labelling was carried out for 5 days at 30-32°C and the RNA-equivalent of 5 g leaf tissue was applied to each slot. The extra band of labelled viroid-RNA is clearly visible. Time of exposure: 10 hrs.

conditions an extra species of ^{32}P-labelled RNA appears in PSTV-infected plants which migrates somewhat slower than the 8S-plant RNA present in both infected and healthy plants. The bioassay of eluates of 2.5 mm gel slices cut according to the radioactive scale present on the left side of the autoradiograph during the exposure showed that the maximum of infectivity was associated with fractions Nr. 33 and 34. These gel slices correspond exactly to the new RNA-band which clearly substantiates that PSTV-RNA must be localized in this band. No infectivity was found in gel slices from the healthy control section of the gel.

The results of a longer exposure of this gel to the X-ray film (24 hrs as compared to 10 hrs in Figure 2) is shown in Figure 5. This procedure allows the detection of several minor species of RNA which appear in a reproducible pattern. Their nature has not been

investigated and it is presumed that they represent nuclear RNA
species.

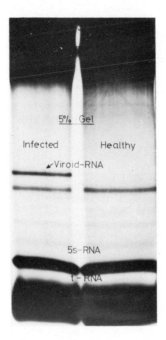

Fig. 5. Autoradiograph of the 5% gel from Figure 4 after 24 hrs of
exposure of the X-ray film to show some minor species of labelled
RNA.

The analysis of a part of the nucleic acid preparation from Figure
5 on a 10% gel is shown in Figure 6. At first sight one might assume
that there is little difference between the RNA pattern from healthy
and infected RNA. But a close examination of the autoradiograph and
the bioassay of gel slices showed that in a 10% gel the viroid-RNA
(see arrow) practically co-migrates with the plant-RNA band. In a
15% and a 20% gel, however, viroid-RNA migrates faster than the 8S-
plant RNA (not shown). This finding substantiates similar reports of
Semancik et al. (1973) and it is evidence for an unusual structure
and conformation of the viroid RNA.

The incorporation studies clearly show that the synthesis of viroid-
RNA is temperature-dependent. At temperatures between 18 and 20°C
only little viroid synthesis occurs whereas at temperatures above
28°C the synthesis of PSTV-RNA in systemically infected tomato plants

242

is considerably increased.

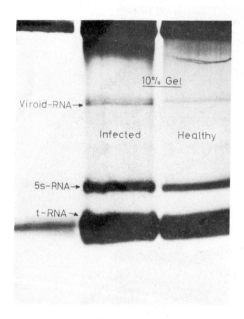

Fig. 6. Autoradiograph of ^{32}P-labelled "2 M LiCl-soluble RNA" from
PSTV-infected and healthy tomato plants on a 10% polyacrylamide
gel. For further details and for comparison see Figures 4 and 5
in which the analysis of aliquots on a 5% gel is shown. It is
interesting to note that in a 10% gel the viroid-RNA practically
co-migrates with the 8S-plant RNA species.

LABELLING OF ExCV-RNA *WITH* ^{32}P.

In a comparative labelling experiment three PSTV-infected tomato
plants and three ExCV-infected *Gynura* plants were exposed to ^{32}P
orthophosphate (2 mCi per plant) for a period of 5 days at 30-32°C.
The analysis of the "2 M LiCl-soluble RNA" on a 5% polyacrylamide gel
is shown in Figure 7. It is evident that both viroids migrate exactly the same distance in the gel which confirms their similarity in
size and conformation. Furthermore, it appeared that under the conditions tested the rate of synthesis of ExCV in *Gynura* is on average
less than half of that of PSTV in tomato, if based on the fresh weight
of the leaf tissue and on the radioactivity incorporated. But one
should bear in mind that both viroid-host plant systems might have
different temperature optima. As judged by biological criteria (see
Table 1) optimal synthesis of ExCV in *Gynura* might, in fact, be

achieved at temperatures above 35°C.

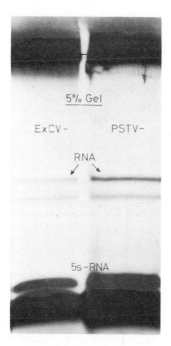

Fig. 7. Autoradiograph of a 5% polyacrylamide gel on which [32]P-
   labelled "2 M LiCl-soluble RNA" from ExCV-infected *Gynura*
   plants and from PSTV-infected tomato plants grown at 30-32°C
   were electrophoresed adjacently.

The "2M LiCl-soluble RNA" from 2 g of leaf tissue was applied
to each slot. The two viroids show an identical pattern of
migration. But, based on the fresh weight of leaf tissue and
on the incorporation of ^{32}P, the rate of synthesis of ExCV-RNA
in *Gynura* appears to be about 20% of that of PSTV in tomato
under identical conditions.

INCORPORATION OF RADIOACTIVE URIDINE.

Since the use of single leaves and of radioactive uridine as precursor would facilitate investigations of viroid-RNA synthesis the
conditions under which viroid-RNA could be labelled *in vivo* and be
detected in gels were studied.

The result of an experiment with PSTV in tomato is shown in Figure
8. Five leaves of about the same age were cut from PSTV-infected
tomatoes 15 days after infection. The plants had been grown at 28°C
and their youngest leaves showed the first typical symptoms of a
systemic viroid-infection. Each leaf was allowed to take up 50 µCi
[3]H-uridine (as described under Material and Methods) and to incorpor-

244

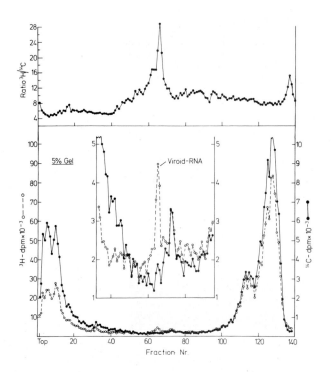

Fig. 8. Analysis of "2 M LiCl-soluble RNA" from 2 g of PSTV-infected (^{3}H-uridine labelled) and from 2 g of healthy (^{14}C-uridine labelled) tomato leaves on a 5% gel.

The precursors were allowed to be taken up by the petiole of individual leaves and to be incorporated for 48 hrs after which the two samples were pooled and processed in one. The inset shows the viroid region of the gel in an extended scale and the upper curve the ratio ^{3}H : ^{14}C.

ate the precursor at 28°C for 48 hrs. Healthy control leaves were exposed to 2.5 μCi ^{14}C-uridine under the same conditions. The experiment was terminated by quick freezing of the pooled leaves from both batches and the RNA was extracted from this pooled sample and prepared for gel-electrophoresis. The analysis of the "LiCl-soluble RNA" from ^{3}H-labelled PSTV-infected leaves and ^{14}C-labelled healthy control leaves is shown in Figure 8. On examination of the two profiles of radioactivity there is little difference at first sight between the RNA from healthy and from PSTV-infected leaves. But there is a minor ^{3}H-peak in fraction 61-64 which becomes evident when this viroid region of the gel is enlarged. The inset in Figure 8 clearly shows that a new species of RNA becomes labelled in PSTV-infected leaves. The comparison with the ^{32}P-autoradiographs from the corresponding slab gels (Fig 5) indicates, that on the basis of its migration this RNA is, in fact, PSTV-RNA. The adjacent faster migrating species of RNA is the 8S-plant RNA occurring in both, healthy and viroid-infected leaf-tissue. This presumption was supported by the result of the infectivity test carried out with a sister gel of the same run. PSTV-infected tomato plants were obtained only with the eluates from the gel slices 60-66, which corresponds exactly to the viroid-peak.

The appearance of the viroid-RNA in infected leaves is also demonstrated by the curve showing the ^{3}H : ^{14}C ratio (Fig 8, top) which has its maximum exactly at the viroid-RNA peak. Furthermore, this curve shows that no additional RNA species are detectable in the 5% gel. Therefore one can assume that no viroid-specific RNAs other than the infectious viroid-RNA molecules accumulate to appreciable concentrations in viroid-infected tissue. Finally, the ratio-curve demonstrates that viroid-RNA synthesis has practically no impact on the synthesis of the soluble RNA species (8S-RNA, 5S-RNA and transfer-RNA) of the cell which are resolved in the 5% gel.

In order to investigate the influence of viroid-RNA synthesis on cellular RNAs of high molecular weight, the corresponding "2 M LiCl-insoluble RNA" was electrophoresed on a composite 2.5% polyacrylamide-0.5% agarose-gel which resolves the DNA and the RNA species of the ribosomes of the cytoplasm and the ribosomes of the chloroplasts. The analysis of the profile of radioactivity (not shown) indicated, that viroid infection has no detectable effect on the synthesis of these high molecular weight nucleic acids.

In the control experiments carried out at a growth temperature of 18-20°C (not shown) no ^{3}H-viroid peak was found under the conditions described above, which is in accordance with the negative result of labelling experiments with ^{32}P carried out at the same temperature with whole plants.

In conclusion, these experiments open one interesting experimental approach to study viroid-synthesis *in vivo* and the results obtained indicate that viroid RNA-synthesis seems to have very little impact on the biosynthesis of the nucleic species of the infected cell.

QUANTITATIVE ASPECTS OF VIROID-RNA SYNTHESIS.

The use of radioactive precursors allows the estimation of viroid-RNA synthesis in relation to the synthesis of cellular RNA. The distribution of radioactivity in different species of RNA obtained after fractionation of "2 M LiCl-soluble RNA" from PSTV-infected tomatoes in 5% polyacrylamide gels is presented in Table 2. The incorporation of ^{32}P into the transfer-RNA and 5S-RNA fraction appears to be little affected by the growth temperature and the 5S-RNA fraction carries about 20% of the label of the transfer-RNA. For unknown reasons, in some experiments this value may reach 40%. Temperature has a significant effect on the incorporation of ^{32}P into the 8S plant RNA which is increased about tenfold at the higher temperature. However, the dramatic effect of growth temperature is on viroid-RNA synthesis. As judged by the incorporation in most cases in these experiments it is increased more than threehundredfold at 30-32°C as compared to the incorporation at a growth temperature of 18-20°C.

The relative concentration of viroid-RNA at 18-20°C, if based on the incorporation of ^{32}P, is less than 0.01% of the transfer RNA but at 30-32°C it increases to about 3%. This value is in good agreement with estimates based on the planimetry of the peak areas obtained after gel electrophoresis of unlabelled RNA where it may range between 2-3.5% of the t-RNA fraction. In most of our RNA-preparations from tomato the t-RNA comprises about 10% of the total RNA. Provided the infected plants are growing at an optimal temperature and provided that they are harvested 3-6 weeks after inoculation the PSTV-RNA may constitute up to 0.3% of the total RNA extracted.

When these estimations were carried out with ExCV-infected *Gynura* it appeared that under our conditions the relative concentration of ExCV-RNA at 30-32°C was only 0.8-1.3% of the t-RNA when all symptom-bearing leaves were harvested and used for extraction two months after inoculation.

ISOLATION OF PURIFIED VIROID-RNA.

The radioactive labelling with ^{32}P facilitates greatly the isolation of viroid-RNA and its final purification. The application of ^{32}P-autoradiography for analytical purposes is shown in Figures 9 and 10 in which the efficiency of an improved isolation method for PSTV-RNA has been checked. The samples taken from different steps of the purification procedure show the presence or absence of the ^{32}P-labelled viroid-RNA and of cellular RNA which permitted the method to be optimized. Furthermore, the use of preparative slab gels and the availability of autoradiographs as precise copies of these gels guarantees the exact location of the different species of RNA after gel electrophoresis without any difficulties. Thus the viroid-RNA band

Table 2

Distribution of ^{32}P radioactivity in fractions of
"2 M LiCl-soluble RNA" from PSTV-infected tomato plants after
electrophoresis in 5% polyacrylamide gels.

RNA fraction	Growth temperature 18-20°C c.p.m.	%	Growth temperature 30-32°C c.p.m.	%
Expt. 1				
4S	11,215,620	100	13,769,739	100
5S	2,383,508	21·25	2,717,149	19·73
8S	18,890	0·17	211,020	1·53
Viroid	1,060[1]	0·009	488,267	3·54
Expt. 2				
4S	15,440,980	100	15,382,970	100
5S	3,522,440	22·81	3,138,211	20·40
8S	26,480	0·16	252,394	1·64
Viroid	1,385[1]	0·009	478,790	3·11

1) No band could be detected in the autoradiograph despite extended exposure.

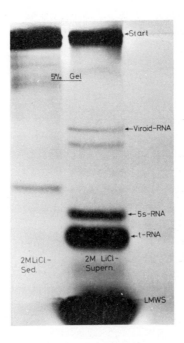

Fig. 9. Autoradiograph of ^{32}P-labelled and 2 M LiCl-fractionated RNA from ExCV-infected *Gynura* plants.

The picture demonstrates the result of an improved LiCl-fractionation (compare with Fig 2 + 3). The ethanol-precipitate of the total RNA was dissolved repeatedly in a solution of 2 M LiCl to isolate the "2 M LiCl-soluble RNA" rather than precipitating out the ribosomal RNA overnight by adding 1 volume of 4 M LiCl to the RNA solution. Viroid-RNA and 8S-plant RNA are found exclusively in the "2 M LiCl-soluble RNA" fraction. LMWS = low molecular weight substances.

may be cut out according to the autoradiograph, avoiding any other plant RNA. After three subsequent extractions of the excised gel bands in 0.1 M TS-buffer pH 7.5 (5 ml buffer per g of gel) usually more than 95% of the total counts in the gel (measured directly in the gel as Cerenkov-counts) were found in the supernatant. The dilute RNA was adsorbed on hydroxyapatite and concentrated as described in Material and Methods.

Figure 11 shows the analysis on a 5% polyacrylamide gel of a sample of PSTV-RNA purified according to this method. The material obtained is practically free from any other contaminating RNA and electrophoretically homogeneous in a 5% gel, as shown by the coinciding peaks of UV-absorbancy and of radioactivity. The comparison of

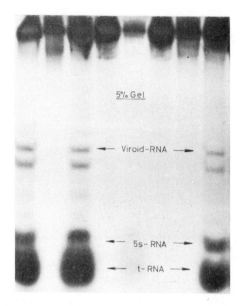

5‰ Gel

←— Viroid-RNA —→

◄— 5s- RNA —►

◄— t- RNA —►

Fig. 10. Autoradiograph of an analytical 5% polyacrylamide gel on
which the efficiency of an isolation method for viroid-(PSTV)RNA
has been checked.

The samples were taken from different steps of the purification
and the presence and absence of the ^{32}P-labelled viroid-RNA
allowed to optimize the procedure. Each sample corresponds to
the RNA-equivalent of 2 g of leaf tissue.

these two viroid-RNA peaks with the UV-absorbancy profile of a mixture
of 5S-RNA and transfer-RNA run in a sister tube shows furthermore,
that the electrophoretic mobility of the PSTV-RNA has not been altered, an indication of its unchanged structural and conformational inte-
grity. The specific activity obtained in tomato in different experiments ranged from 1-5 x 10^5 ^{32}P-c.p.m. (Cerenkov) for 1 µg of purified
PSTV-RNA. The specific activities for 1 µg ExCV purified from
Gynura was usually 5-8 x 10^4 ^{32}P-c.p.m. (Cerenkov). The amount of
^{32}P incorporated into the viroid-RNA meets the requirements for biochemical studies on viroid structure.

DISCUSSION

Radioactive labelling of viroid-RNA presents unique problems as
compared to the labelling of cellular and viral RNA species. Viroid
replication appears to be inordinately temperature-dependent in a way
that differs strikingly from the replication of conventional plant
viruses. Viroid synthesis seems to be best at greenhouse temperatures
which are usually neither tolerated by plant virus systems nor by the

250

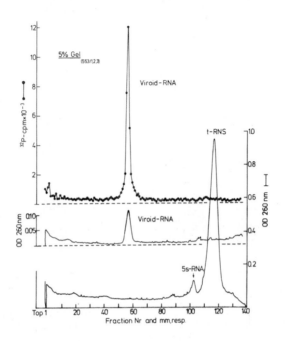

Fig. 11. Analysis of a sample of purified ^{32}P-labelled PSTV-RNA on a 5% polyacrylamide gel.

The RNA was extracted from an excised band from a preparative gel according to an autoradiograph and further purified by chromatography on hydroxyapatite. A mixture of 5S-RNA and t-RNA was run in a sister tube for comparison.

horticulturist involved. Accordingly, the isolation and purification of viroid-RNA may fail if the host plants in which these agents are propagated are grown under conventional conditions like most plants for normal plant virus work. Thus, our previous failure to demonstrate viroid-RNA as a distinct UV-absorbing peak even in our most concentrated preparations, was primarily the result of growing our viroid-infected plants at "normal" greenhouse temperature. In our climate this ranges between 18 and 22°C most time during the year and at this temperature viroid synthesis - and accumulation, is rather poor. The concentrations obtained may be sufficient for detection by bioassay but they are by far too low for the spectrophotometric location in analytical polyacrylamide gels. Raising the growth temperature from 18-20°C to 30-32°C may cause a more than three-hundredfold increase in viroid yield. This finding explains why the successful isolation and demonstration of viroid RNA as a distinct UV-absorbing peak was first reported from laboratories where comparatively high greenhouse temperatures are provided most time of the year by the climate itself (Diener, 1972; Semancik and Weathers, 1972).

The most plausible explanation for the increased viroid synthesis at temperatures above 30°C might lie in the fact that the viroids known at present are restricted to host plants growing under subtropical and tropical conditions (citrus exocortis viroid) in continental climates (potato spindle tuber viroid) or in greenhouses, like the chrysanthemum stunt viroid (Diener and Lawson, 1973) and the cucumber pale fruit viroid (van Dorst and Peters, 1974). Accordingly, the "relative temperature-dependence" of viroid synthesis might be primarily a matter of adaptation selected for during viroid-evolution in their hosts. One is tempted to speculate, that this behaviour might be related to some regulatory mechanism acting at higher temperatures in favour of viroid-RNA replication.

Unfortunately the "long term" labelling experiments described here for systemically infected whole plants and individual leaves do not allow any statements on the actual mode of viroid replication for which actual pulse-labelling is needed. Thus, it cannot be decided, as yet, whether only a small number of cells exhibits a high rate of viroid-synthesis or whether viroid-RNA is synthesised at a low rate in all or almost all cells at the same time. Consequently, the failure to detect any direct competition of viroid synthesis with the RNA metabolism of the host cell must not be overemphasized. For similar reasons the use of different inhibitors of nucleic acid- and protein-biosynthesis for the elucidation of viroid replication in intact plant tissue has severe limitations. Irrespective of these problems, the successful radioactive labelling of viroid RNA will help to resolve certain questions of viroid-structure and -function.

ACKNOWLEDGEMENTS

This study was supported by the Sonderforschungsbereich 47 (Virology). We thank Dr. E.L. Sattler for his help in the early stages of this work.

REFERENCES

Diener, T.O. (1971). Potato spindle tuber "virus". IV. A replicating, low molecular weight RNA. *Virology*. 45, 411-428.

Diener, T.O. (1972). Potato spindle tuber viroid. VIII. Correlation of infectivity with a UV-absorbing component and thermal de-naturation properties of the RNA. *Virology*. 50, 606-609.

Diener, T.O., and Lawson, R.H. (1973). Chrysanthemum stunt: A viroid disease. *Virology*. 51, 94-101.

Diener, T.O., and Raymer, W.B. (1969). Potato spindle tuber virus: A plant virus with properties of a free nucleic acid. II. Charac-terization and partial purification. *Virology*. 37, 351-366.

Diener, T.O., and Smith, D.R. (1971). Potato spindle tuber viroid. VI. Monodisperse distribution after electrophoresis in 20% poly-acrylamide gels. *Virology*. 46, 498-499.

Haviland, R.T., and Bieber, L.L. (1970). Scintillation counting of ^{32}P without added scintillator in aqueous solutions and organic solvents and on dry chromatographic media. *Analytical Biochemistry*. 33, 323-334.

Peacock, A.C., and Dingman, C.W. (1967). Resolution of multiple ribo-nucleic acid species by polyacrylamide gel electrophoresis. *Bio-chemistry*. 6, 1818-1827.

Raymer, W.B., and Diener, T.O. (1969). Potato spindle tuber virus: A plant virus with properties of a free nucleic acid. I. Assay, extraction, and concentration. *Virology*. 37, 343-350.

Sänger, H.L. (1972). An infectious and replicating RNA of low molecular weight: The agent of the exocortis disease of citrus. *Advances in Biosciences*. 8, 103-116.

Semancik, J.S., Magnuson, D.S., and Weathers, L.G. (1973). Potato spindle tuber disease produced by pathogenic RNA from citrus exocortis disease: Evidence for the identity of the causal agents. *Virology*. 52, 292-294.

Semancik, J.S., and Weathers, L.G. (1972). Exocortis disease: Evi-dence for a new species of "infectious" low molecular weight RNA in plants. *Nature New Biology*. Vol. 237.

Van Dorst, H.J.M., and Peters, D. (1974). Some biological observations on pale fruit, a viroid-incited disease of cucumber. *Netherlands Journal of Plant Pathology*. 80, 85-96.

CHARACTERIZATION OF THE COMPLEX FORMED BETWEEN PS8 cRNA AND DNA ISOLATED FROM A_6-INDUCED STERILE CROWN GALL TISSUE

R.A. SCHILPEROORT, J.J.M. DONS AND H. RAS

*Biochemisch Laboratorium,
Rijksuniversiteit Leiden, The Netherlands.*

Different features of the hybridization reaction between crown gall DNA and cRNA transcribed from PS8 DNA and A_6 DNA were investigated. The hybrids formed have been characterized with regard to RNAase and DNAase resistance, melting temperature, S-value of the cRNA melted from the RNAase treated hybrids, and buoyant density in Cs_2SO_4 gradients. All the data indicate that the cRNA molecules do not anneal to homologous DNA present in the crown gall DNA preparations, but possibly to RNA molecules present in crown gall DNA preparations. The reaction of PS8 cRNA with crown gall DNA freed from the hybridizing component shows that the amplification of PS8 genomes in crown gall DNA, as supposed earlier, does not exist. The possible origin of these molecules in relation to the plasmid hypothesis of tumour induction is discussed.

The plant tumour called crown gall arises under the influence of virulent *Agrobacterium tumefaciens* cells inoculated into wounded dicotyledonous plants. Crown gall formation is believed to be a real neoplastic disease having several characteristics in common with animal tumours. For this reason, and also because of the fact that the disease might represent a naturally occurring case of genetic information transfer from the bacterium into plant cells (Chadha and Srivastava, 1971; Goldman-Ménage, 1970; Lejeune, 1972; Milo and Srivastava, 1969; Quétier *et al.*, 1969; Schilperoort, 1969, 1971; Schilperoort *et al.*, 1967, 1969, 1973; Sittert, 1972; Srivastava, 1970; Torok and Cornesky, 1970), the crown gall problem is an interesting subject for study.

A lysogenic phage of *A.tumefaciens* named PS8 has frequently been related with the tumour inducing capacity of *A.tumefaciens*. Conflicting data about its possible role in tumour induction have been published (Tourneur and Morel, 1970; Leff and Beardsley, 1970; Beiderbeck *et al.*, 1973). In any case, the induction of PS8 is not likely to be an essential prerequisite since non-phase-producing strains were found

to be as pathogenic as their PS8-lysogenic derivatives showing a high
frequency of spontaneous induction (Brunner and Pootjes, 1969). Recently, however, base-sequence homology has been detected between PS8
cRNA and DNA from sterile crown gall tissue induced by strain A6
(Schilperoort *et al.*, 1973). The interpretation of this result is
complicated by the fact that no PS8 can be induced from the PS8-insen-
sitive A_6 cells. Moreover, no PS8-DNA can be detected in DNA isolated
from these A_6 cells. It has been postulated that a cryptic PS8 is
integrated in a large plasmid found to be present in A_6 cells which is
easily lost with routine DNA isolation procedures. During tumour induction this plasmid is then transferred into the plant cell after
bacterial attachment to the cell wall (Schilperoort, 1969).

Very important progress in crown gall research has been made recently by the finding of an absolute correlation between the presence
of a large plasmid and virulence of *A.tumefaciens* strains (Zaenen
et al., 1974). The hypothesis has been made that this type of plasmid
is the Tumour Inducing Principle (TIP).

Here we report studies on the features of the hybridization reaction
between crown gall DNA and PS8 cRNA or A_6 cRNA and on the characterization of the complexes formed.

RESULTS

THE HYBRIDIZATION REACTION.

It was already found (Schilperoort, *et al.*, 1973) that DNA from
sterile crown gall tissue and to a much smaller extent also DNA from
normal callus tissues reacted with PS8 cRNA and A_6 cRNA. No reaction
was ever found with tobacco leaf DNA, stem DNA and calf thymus DNA.
Only crown gall DNA could discriminate between the cRNA's used. It
was therefore concluded that the observed reaction with these cRNA's
is specific for DNA isolated from tissue cultured *in vitro* while in
addition crown gall DNA might contain exogenous base sequences not
present in normal callus tissue. RNAase resistant hybrids are found
for both DNA baked on to filters and DNA in solution using the liquid
hybridization method (Schilperoort *et al.*, 1967). The RNAase resistance is about 75%.

Under the same reaction conditions of 2 x SSC for 22 hrs at $66^{\circ}C$
no reaction is found with T4 cRNA (Table 1). The negative result
obtained with T4 cRNA is not due to an absence of the crown gall component otherwise reacting with PS8 cRNA, since the same T4 DNA filter
gave a positive reaction with PS8 cRNA on a subsequent incubation.

Table 1
Hybridization with 10 µg DNA baked
to Millipore filters (0.45 µm) and
PS8 cRNA (465,000 cpm/µg) and T4 cRNA (61,000 cpm/µg)
in the presence of E.coli rRNA.

Reaction conditions: 20 hrs at 66°C in 2 x SSC.

DNA	cRNA	cpm on filter Before RNAase	After RNAase
PS8	PS8 2 µg	413,150	270,800
Crown gall	PS8 2 µg	19,282	14,669
Blank filter	PS8 2 µg	172	75
T4	T4 1 µg	7,066	3,773
Crown gall	T4 2 µg	164	
	PS8 2 µg*	13,708	10,257
Blank filter	T4 2 µg	75	
	PS8 2 µg*	746	371

* After hybridization with T4 cRNA the same filter is incubated
with PS8 cRNA.

To see whether RNA with a higher GC content or a mixture of all
the different types of RNA normally present in tobacco leaf cells
could influence the reaction of crown gall DNA with either PS8 cRNA
or A_6 cRNA we performed competition hybridization experiments with
melted E.coli rRNA and melted tobacco leaf total RNA. The reaction of
A_6-[3]H cRNA was not influenced by either type of RNA up to an amount of
300 µg RNA in the reaction mixture. The results of the reaction between crown gall DNA and PS8-[3]H cRNA in the presence of various
amounts of melted E.coli rRNA is shown in Table 2. In this case also,
no influence of the competitive RNA is detected. For this reason we
routinely add about 200 µg E.coli rRNA to the reaction mixture, since
this reduces background noise considerably and thus enhances the specificity of the reaction. These results show that crown gall DNA has
a preference for PS8 cRNA and A_6 cRNA and does not simply react with
any RNA.

The kinetics of the reaction of crown gall DNA, normal callus DNA
and A_6 DNA with A_6 cRNA is shown in Figure 1. The influence of incubation temperature on the amount of hybrid formed between crown gall
DNA and PS8 cRNA is shown in Table 3.

Table 2

The influence of *E.coli* rRNA

on the amount of hybridization of 10 μg crown gall DNA

with 1.8 μg PS8 cRNA (465,000 cpm/μg) and on

the background noise of empty filters.

DNA	*E.coli* rRNA	Filter bound cpm after RNAase	% of input cpm
Crown gall	--	18,408	2.20
Crown gall	100 μg	23,922	2.86
Crown gall	200 μg	18,830	2.27
Blank filter	--	2,610	0.31
Blank filter	100 μg	405	0.05
Blank filter	200 μg	185	0.02

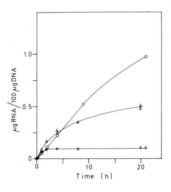

Fig. 1. Kinetics of the hybridization of 3 μg crown gall DNA (Δ——Δ) and 3 μg normal tissue DNA (x——x) with a saturating amount *A.tumefaciens* A_6 ^{3}H-cRNA (3 μg) and of the hybridization of 5 μg *A.tumefaciens* A_6 DNA (o——o) with 1 μg *A.tumefaciens* A_6 ^{3}H-cRNA.

A lower temperature or a shorter time of incubation decrease the amount of hybrid formed. These data exclude the possibility that the observed reaction is of a simple, nonspecific kind, but suggest a reaction between complex cRNA molecules and other complex molecules, most probably having a polynucleotide nature.

Table 3

The influence of temperature on the amount of
hybridization between 10 µg filter bound crown gall DNA
and 1.8 µg PS8 cRNA (465,000 cpm/µg).

	cpm on filter	
	Before RNAase	After RNAase
22 hrs 66°C	18,766	13,014
22 hrs 50°C	8,221	3,680
22 hrs room temperature	1,829	1,267

REPRODUCIBILITY OF THE EXPERIMENTS.

The detection of the complex between crown gall DNA and PS8 cRNA
depends on the procedure used to isolate tumour DNA and on the batch
of Millipore filter sheet from which 22 mm filters are made. When
DNA is isolated using a Sepharose 4B column in 2 M NaCl (Heijn *et al.*,
1973) it almost completely looses the component which is able to react with PS8 cRNA, otherwise present in DNA isolated only with a SDS-
pronase-phenol procedure (Schilperoort, 1969) (Table 4). The same
occurs when the DNA is chromatographed on a hydroxylapatite (HAP)
column and the double-stranded DNA fraction is taken for hybridization with PS8 cRNA. HAP chromatography completely removes the polysaccharides which are always present in variable amounts in crown gall
DNA. Polysaccharides are not completely removed by gel filtration
through Sepharose 4B.

Table 4

Dependance of the hybridization reaction
between crown gall DNA and PS8 cRNA on the isolation
and purification method of the DNA.

Crown gall DNA	cpm on filter	
	Before RNAase	After RNAase
SDS-Pronase-Phenol isolation	19,282	14,669
Sepharose-4B isolation	559	284
Hydroxylapatite purification	520	175

The gel filtration procedure very efficiently frees DNA from RNA and proteins as has been described (Heijn et al., 1973). A polysaccharide is not likely to be the reactive component in crown gall DNA in its reaction with PS8 cRNA, since it was found that the polysaccharides are completely eliminated when the denatured DNA is adsorbed on to filters and the filters washed afterwards. The total amount of polysaccharide present in the DNA preparation is found in the filtrates. Also a protein is presumably not the reactive component since an extensive additional pronase treatment of crown gall DNA did not significantly change its capacity to hybridize with PS8 cRNA.

Rather surprisingly we latterly found that some unknown quality of the Millipore filters determines whether the complex of crown gall DNA with PS8 cRNA can be detected or not. This phenomenon has not been detected previously. Depending on the batch of Millipore filters, either a large amount of complex or only 5-10% of its amount is detected reproducibly even when one and the same crown gall preparation is used (Table 5).

Table 5

The amount of hybridization of PS8 cRNA
with crown gall DNA and PS8 DNA on two different batches of
Millipore HA filters (A and B).

DNA	Filter	cpm on filter Before RNAase	After RNAase
Crown gall 10 µg	A	19,282	14,669
Crown gall 10 µg	B	1,441	994
PS8 0.1 µg	B	29,144	18,644
PS8 0.1 µg and Crown gall 10 µg	B	26,330	17,673
PS8 10 µg	A	473,435	364,060
PS8 10 µg	B	409,068	304,860

The two different types of filter will henceforth be called "positive" and "negative" filters. The DNA binding ability seems to be the same for both types of filter, since the amount of DNA that remains fixed to the filters after hybridization is almost the same. Also the reaction of PS8 cRNA with PS8 DNA behaves normally even when the PS8 DNA is mixed with crown gall DNA containing large amounts of polysaccharides to see whether the PS8 signal is masked in some way when using the "negative" filter type (Table 5). The "negative" results obtained with several batches of Millipore filters is not specific for this make since "negative" results were also found for filters from Schleicher and Schüll and Sartorius having the same pore diameter (0.45 μ).

All these data suggest that the component in crown gall DNA which rather specifically reacts with PS8 cRNA can hardly be simply PS8 DNA.

CHARACTERIZATION OF THE CROWN GALL DNA-PS8 cRNA COMPLEX.

a. Melting temperature (Tm).

The Tm in 2 SSC of the PS8 DNA-PS8 cRNA hybrid is $87^{\circ}C$. The melting profile is not very much changed when the hybrids are treated with RNAase A+Tl (Fig 2). The melting profiles of the hybrids between crown gall DNA with PS8 cRNA also show a narrow transition width when the hybrids are not treated with RNAase A+Tl (Fig 2). The Tm, however, is lower than found for the homologous reaction of the cRNA and is $80^{\circ}C$. The narrow transition width suggests that we are dealing with a very regular structure, like hybrids formed between complementary polynucleotide chains. Nonspecific binding of cRNA gives a clearly different melting profile and can be directly recognized. This is shown for cRNA adsorbed to empty filters and for some dirty complexes sometimes formed after liquid hybridization of crown gall DNA with PS8 cRNA giving incredible numbers of counts (Fig 2).

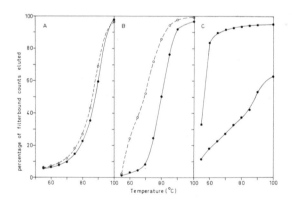

Fig. 2. Melting curves of (A). PS8 DNA hybridized with PS8 ^{3}H-cRNA, RNAase treated (o———o) and without RNAase treatment (●———●); (B). crown gall DNA hybridized with PS8 ^{3}H-cRNA, RNAase treated (o———o) and without RNAase treatment (●———●); (C). empty filter after incubation with PS8 ^{3}H-cRNA (●———●) and crown gall DNA after liquid hybridization with PS8 ^{3}H-cRNA and binding to Millipore filter, RNAase treated (■———■).

Unexpectedly the melting profile changes considerably when the crown gall DNA-PS8 cRNA hybrids are treated with RNAase A+T1 (Fig 2) although the RNAase resistance (73%) of these hybrids is completely comparable to that of the PS8DNA-PS8cRNA hybrids.

b. S-value of RNA melted from RNAase treated crown gall DNA cRNA hybrids

If mis-matching has occurred along extensive stretches of base-sequences one would expect that, after RNAase treatment of the hybrids, the RNA molecules melted from the hybrids should have a much lower S-value than the input RNA. We found, however, that the major fraction of the RNA molecules melted from the crown gall-cRNA hybrids which were RNAase treated have a mean S-value of about 10 in a 5-20% sucrose gradient, while only a minor fraction of the molecules indeed sedimented heterogeneously at lower values (Fig 3). This is found for both the hybrids with A_6 cRNA and PS8 cRNA. The mean S-value of the RNA molecules melted from the hybrids is higher than the mean S-value of the cRNA molecules used for the hybridization reaction. The largest RNA molecules have therefore reacted with crown gall DNA and must have fairly well matched with the reactive component in the DNA preparation.

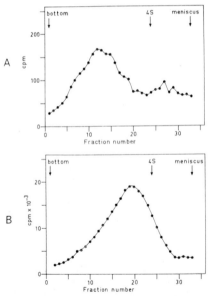

Fig. 3. Sedimentation profile of: (A). *A.tumefaciens* A_6 ^{3}H-cRNA (heated for 5 min at 100°C); (B). *A.tumefaciens* A_6 ^{3}H-cRNA, melted from a crown gall DNA - *A.tumefaciens* A_6 ^{3}H-cRNA hybrid after 18 hrs at 41,000 rpm in a 5%-20% sucrose gradient.

c. Buoyant density of crown gall DNA and its hybrid with PS8 cRNA in
Cs_2SO_4 *gradients.*

From Figure 4 it can be seen that the RNAase treated hybrid of
crown gall DNA with PS8 cRNA has a density of 1.62 g/ml. Such a
fraction is not found for either tobacco leaf DNA or PS8 DNA. It is
important to note that the homologous PS8 DNA-PS8 cRNA hybrid bands
close to DNA. The crown gall-PS8 hybrid has the same density as
double-stranded RNA. Double-stranded RNA is formed during hybridization because of the presence of self-annealing cRNA molecules which
are always present in variable amounts in the PS8 cRNA preparations.
The crown gall-PS8 hybrid can be discriminated from the double-stranded cRNA molecules because of its unique property of adsorbing to
nitrocellulose filters; cRNA does not bind (Fig 5). Also, on negative filters these crown gall DNA-PS8 cRNA hybrids can still be detected, although in much lower amount.

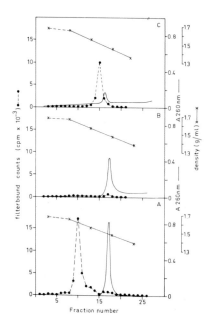

Fig. 4. Analysis of RNAase treated hybridization mixtures by Cs_2SO_4
gradients. A260 was recorded by passage through a capillary
flow cell (light path 2 mm) and subsequently fractionated. Fractions were diluted with 6 x SSC and filtered through Millipore
filters. (A). crown gall DNA-PS8-^{3}H-cRNA; (B). tobacco leaf
DNA-PS8 ^{3}H-cRNA; (C). PS8 DNA-PS8 ^{3}H-cRNA.

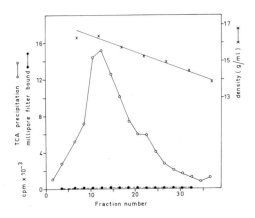

Fig. 5. Analysis in Cs_2SO_4 gradient of RNAase treated PS8-cRNA after incubation for 20 hr in 2 x SSC at 66°C. Total number of TCA precipitable counts and Millipore filter bound counts.

In contrast to the PS8DNA-PS8 cRNA hybrids, the hybrids formed between crown gall DNA and PS8 cRNA isolated from Cs_2SO_4 gradients are not sensitive to DNAase treatment. Furthermore, not denaturing the DNA or hydrolysis of the DNA by DNAase before hybridizing it to cRNA has no influence on the amount of hybrid formed (Table 6).

Table 6

Reaction of 10 µg single-stranded,
double-stranded and degraded DNA with PS8 cRNA (1.8 µg).

DNA	DNA treatment	cpm on filter Before RNAase	After RNAase
Crown gall	--	26,524	19,740
Crown gall	DNAase	27,138	17,891
PS8	--	502,317	375,053
PS8	DNAase	262	123
Crown gall	--	18,928	13,014
Crown gall	Not melted	18,050	12,921
PS8[*]	--	157,426	104,694
PS8[*]	Not melted	2,740	1,570

* filters with 1 µg DNA.

To see whether the component in crown gall DNA reacting with PS8 cRNA itself has the density of about 1.6 g/ml, denatured DNA was fractionated in a Cs_2SO_4 gradient. The fractions were baked on to Millipore filters and subsequently hybridized with PS8 cRNA. Unfortunately only "negative" filters were available. Nevertheless from Figure 6 it can be seen that fractions with an RNA density give a positive reaction with PS8 cRNA. Such a fraction is not found for tobacco leaf DNA.

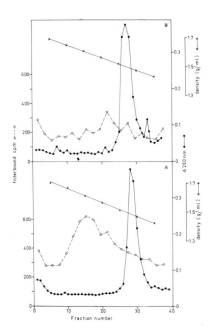

Fig. 6. Analysis of crown gall DNA (A) and tobacco leaf DNA (B) in Cs_2SO_4 gradients. 0.2 ml fractions were diluted with 0.5 ml 6 X SSC, the A260 was determined, and subsequently the fractions were filtered through Millipore filters. Each filter was hybridized with 1 µg PS8 ^{3}H-cRNA and RNAase treated.

All the data together suggest that the observed hybridization of crown gall DNA with PS8 cRNA is not due to DNA. The buoyant density of the hybrid and the reacting component in the DNA sample seems to indicate the presence of RNA in the DNA preparation. Since it is known that RNA molecules can adsorb to nitrocellulose filters only if covalently bound to a poly A tract, we tested whether poly A containing molecules were present in our crown gall DNA preparations. DNA

was isolated from crown gall tissue labelled for 3 hrs with 2 mCi of
[3]H-adenosine.   Indeed a small labelled compound with filter binding
capacity was found at the RNA density;   negative filters were used
(Fig 7).   However, preliminary experiments showed that those fractions which hybridized with PS8 cRNA after filter binding did so after
an 18 hr treatment with 0.3 N NaOH at 37°C, indicating that RNA might
not be the hybridizing component.

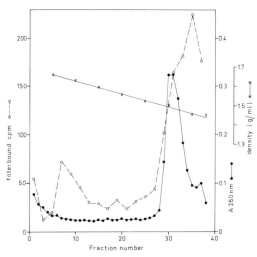

Fig. 7. Cs_2SO_4 gradient analysis of DNA isolated from ^{3}H-adenosine
labelled crown gall tissue. Fractions were diluted with 6 x SSC,
A260 determined and filtered through Millipore filters.

DISCUSSION

From the data shown it is clear that the results from filter
hybridization experiments between crown gall DNA and PS8 cRNA have to
be interpreted in a different way.   Although the presence of a small
amount of PS8 DNA in crown gall DNA cannot be excluded, this DNA is
not detected by these filter hybridization methods.   The DNAase resistance, independence of denaturation, and buoyant density in
Cs_2SO_4 gradients of the hybridizing component show that it can hardly
be DNA.   The presence of a small amount of RNA in the crown gall DNA
preparations and the banding of the crown gall-PS8 cRNA hybrid at RNA
density suggest that RNA is the hybridizing component.   The observed
capacity of the RNA to adsorb to nitrocellulose filters can be explained by assuming that the RNA molecules are covalently bound to a
poly A tract.   It is a very well known fact today that a poly A tract

bound to mRNA molecules causes these molecules to bind to nitrocellulose (Sullivan and Roberts, 1973). Furthermore, this binding is highly dependent upon the quality of the filters used. All other types of RNA do not adsorb, which is also shown in our experiments where PS8 cRNA on filtration is not retained by the filters. Although the density of the hybrid and the hybridizing component suggest the importance of RNA in the hybridization reaction, the apparent alkali resistance of this component is in contradiction with this.

Now that we are able to isolate the molecules in crown gall DNA preparations reacting with PS8 cRNA an accurate analysis can be made. More should be known about the specificity towards PS8 cRNA by comparing hybridization with other cRNA's and about the specificity for crown gall cells. The hybridizing compound is not found in DNA from tissues of healthy plants like leaves and stems. However, DNA from normal callus tissue grown *in vitro* also reacts with PS8 cRNA to a lower extent. The "normal" callus tissue that has been used is able to grow, like crown gall tissue, without the addition of phytohormones (Dons *et al.*, 1974).

The reaction of PS8 cRNA with crown gall DNA shows that the amplification of PS8 genomes, as supposed earlier (Schilperoort *et al.*, 1973), does not exist. We even doubt whether a complete PS8 genome is present in the genome of the crown gall cells, since in preliminary DNA renaturation experiments (C_0t technique using HAP chromatography) an amount of PS8 DNA equivalent to one PS8 genome was not detected. (T. Currier, unpublished). The same holds for DNA isolated by a routine method (Schilperoort, 1969) from A_6 cells.

However, several lines of evidence indicate that a plasmid present in cells of virulent strains and not in cells in wild type avirulent strains (Zaenen *et al.*, 1974) is directly involved in tumour induction. Recently it was found that a virulent strain that could be converted to avirulence simply by culturing at 37°C (Hamilton and Fall, 1971) concomitantly lost its plasmid (Larebeke *et al.*, 1974). Moreover, we found that the avirulent derivative was no longer able to make use of nopaline as a source of nitrogen. It has been stated that the arginine derivatives octopine and nopaline are present in crown gall cells only (Goldman-Ménagé, 1970; Lejeune, 1972). The presence of one or the other amino acid derivative was found to be independent of the plant species used but completely dependent upon the type of bacterial strain used for transformation. Although others

(Johnson *et al.*, 1974; Wendt-Gallitelli and Dobrigheit, 1973) deny the specificity of these guanidine derivatives for crown gall cells, a detailed analysis of this phenomenon in our laboratory (manuscript in preparation) completely confirms the earlier results. In our opinion a plasmid carrying the information for either octopine or nopaline is transferred after bacterial attachment into the plant cells in the period of crown gall induction.

If on further analysis the hybridizing component in crown gall DNA turns out to be RNA complementary to PS8 cRNA then, in the light of the plasmid hypothesis of tumour induction, these molecules may come from only a few genes existing in both the phage genome and the plasmid of *A.tumefaciens* strains even when these strains are insensitive to PS8 and non-lysogenic for PS8. In the process of tumour induction these genes might facilitate the integration of part or the complete plasmid. Actually we do not know yet whether the plasmid or a part of it is integrated in the host genome. Experiments with plasmid DNA and plasmid cRNA will therefore be performed to determine its presence in crown gall DNA and to see whether there is a correlation with the hybrids described here.

ACKNOWLEDGEMENTS
We gratefully acknowledge help and discussion of Professor E.W. Nester and T. Currier (Department of Microbiology, University of Washington, Seattle, U.S.A.) during their stay in our laboratory.

REFERENCES
Beiderbeck, R., Heberlein, G.T., and Lippincott, J.A. (1973). On the question of crown gall tumour initiation by DNA of bacteriophage PS8. *Journal of Virology.* 11, 345-350.

Brunner, M., and Pootjes, C.F. (1969). Bacteriophage release in a lysogenic strain of *Agrobacterium tumefaciens*. *Journal of Virology.* 3, 181-186.

Chadha, K.C., and Srivastava, B.I.S. (1971). Evidence for the presence of bacteria-specific proteins in sterile crown gall tumour tissue. *Plant Physiology.* 48, 125-129.

Dons, J.J.M., Valentijn, M., Schilperoort, R.A., and van Duijn, P. (1974). Nuclear DNA content and phytohormone requirements of normal, crown gall and habituated tissues of *Nicotiana tabacum* var. White Burley. *Experimental Cell Research*, in press.

Goldman-Ménagé, A. (1970). Recherches sur le métabolisme azoté des tissus de crown gall cultivés "in vitro". *Thesis. Annales des Sciences Naturelles Botanique.* 12, 223-310.

Hamilton, R.H., and Fall, M.Z. (1971). The loss of tumour-initiating ability in *A.tumefaciens* by incubation at high temperature. *Experientia.* 27, 229-230.

Heijn, R.F., Hermans, A.K., and Schilperoort, R.A. (1973). Rapid and efficient isolation of highly polymerized plant DNA. *Plant Science Letters.* 2, 73-78.

Johnson, R., Guderian, R.H., Eden, F., Chilton, M.D., Gordon, M.P., and Nester, E.W. (1974). Detection and quantitation of octopine in normal plant tissue and in crown gall tumours. *Proceedings of the National Academy of Sciences, U.S.A.* 71, 536-539.

Larebeke, N., van, Engler, G., Holsters, M., van den Elsacker, S., Zaenen, I., Schilperoort, R.A., and Schell, J. (1974). Direct evidence indicating that the presence of a large plasmid in *Agrobacterium tumefaciens* strains is essential to the crown gall inducing ability of such strains. *Nature*, in press.

Leff, J., and Beardsley, R.E. (1970). Action tumorigène de l'acide nucléique d'un bactériophage présent dans les cultures de tissu tumoral de Tournesol *(Helianthus annuus). Compte rendue hebdomadaire des séances de l'Académie des Sciences.* 270, 2505-2507.

Lejeune, B. (1972). Recherches sur le métabolisme de la lysopine dans les tissue de crown gall de scorsonère cultivés *in vitro. Thesis, Paris.*

Milo, G.E., and Srivastava, B.I.S. (1969). RNA-DNA hybridization studies with the crown gall bacteria and the tobacco tumour tissue. *Biochemical and Biophysical Research Communications.* 34, 196-199.

Quétier, F., Huguet, T., and Guillé, E. (1969). Induction of crown gall: Partial Homology between tumour-cell DNA, Bacterial DNA and the G + C-rich DNA of stressed normal cells. *Biochemical and Biophysical Research Communications.* 34, 128-133.

Schilperoort, R.A. (1969). Investigations on plant tumours. Crown gall. On the biochemistry of tumour induction by *Agrobacterium tumefaciens. Thesis, Leiden.*

Schilperoort, R.A. (1971). Integration of *Agrobacterium tumefaciens* DNA in the genome of crown gall tumour cells and its expression. *Proceedings of the Third International Conference on Plant Pathogenic Bacteria, Wageningen, 1971 (Pudoc, Wageningen).* 223-238.

Schilperoort, R.A., Veldstra, H., Warnaar, S.O., Mulder, G., and Cohen, J.A. (1967). Formation of complexes between DNA isolated from tobacco crown gall tumours and RNA complementary to *Agrobacterium tumefaciens* DNA. *Biochimica et Biophysica Acta.* 145, 523-525.

Schilperoort, R.A., Meijs, W.H., Pippel, G.M.W., and Veldstra, H. (1969). *Agrobacterium tumefaciens* cross-reacting antigens in sterile crown gall tumours. *F.E.B.S. Letters.* 3, 173-176.

Schilperoort, R.A., van Sittert, N.A., and Schell, J. (1973). The presence of both PS8 and *Agrobacterium tumefaciens* A_6 DNA base sequences in A_6-induced sterile crown gall tissue cultured *in vitro*. *European Journal of Biochemistry.* 33, 1-7.

Sittert, N.J. van (1972). Investigations on plant tumours. Analysis of DNA and RNA in the crown gall tumour cell. *Thesis, Leiden.*

Srivastava, B.I.S. (1970). DNA-DNA hybridization studies between bacterial DNA, crown gall tumour cell DNA and the normal cell DNA. *Life Sciences.* 9, 889-892.

Sullivan, N., and Roberts, W.K. (1973). Characterization and poly(adenylic acid) content of *Ehrlich Ascites* Cell Ribonucleic Acids fractionated on unmodified cellulose columns. *Biochemistry.* 12, 2395-2403.

Torok, D. de, and Cornesky, R.A. (1970). Antigenic and immunological determinants of oncogenesis in plants infected with *Agrobacterium tumefaciens*. *Colloques internationeaux C.N.R.S. no. 193, Les cultures de tissus de plantes.* 443-451.

Tourneur, J., and Morel, G. (1970). Sur la présence de phages dans les tissus de "crown gall" cultivés *in vitro*. *Compte rendu hebdomadaire des séances de l'Académie des Sciences.* 270, 2810-2812.

Wendt-Gallitelli, M.F., and Dobrigheit, I. (1973). Investigations implying the invalidity of octopine as a marker for transformation by *Agrobacterium tumefaciens*. *Zeitschrift für Naturforschung.* 28, 768-771.

Zaenen, I., van Larebeke, N., Teuchy, H., and Schell, J. (1974). Supercoiled circular DNA in crown-gall inducing *Agrobacterium* strains. *Journal of Molecular Biology.* 86, 109-127.

EXPERIMENTS ON THE HOMOLOGY BETWEEN DNA FROM
AGROBACTERIUM TUMEFACIENS AND FROM CROWN GALL TUMOUR CELLS

U.E. LOENING, D.N. BUTCHER,[*] W. SCHUCH
AND M.L. SARTIRANA

*Department of Zoology, University of Edinburgh,
West Mains Road, Edinburgh, EH9 3JT.*

RNA complementary to DNA from *Agrobacterium tumefaciens*
was hybridised to the DNA of tumour and control callus
tissues. The proportion of the plant DNA hybridised was
equivalent to 3-5 Agrobacterium genomes per plant genome,
for both tumour and control. The hybrids formed had
melting temperatures about 7^0 lower than that of the
homologous hybrid with the bacterial DNA. Since no perfectly matched hybrid specific to tumour cell DNA could
be detected, we conclude that Agrobacterium DNA is not
present or integrated into the plant genome.

It is attractive to think that there is an underlying mechanism
for the induction of tumour growth in both plants and animals, and
that this mechanism involves some integration of a foreign DNA into
the induced tumour cell. · In mammalian tissues, it is known that
polyoma or its DNA can induce tumours and that SV40 or its DNA can
likewise transform cells which may then become tumour cells (Dulbecco,
1969; Green, 1970). The more recent finding that the oncogenic RNA
animal viruses may also cause tumours by the synthesis of DNA which
becomes incorporated into the host genome, provides an additional
point to the involvement of nucleic acid in the tumour inducing process (Spiegelman *et al.*, 1970; Temin and Mizutani, 1970). The possibility has therefore been suggested that the DNA of *Agrobacterium
tumefaciens* or perhaps of a phage that it may contain, is the tumour
inducing principle for the formation of crown galls in plants (Srivastava and Chadha, 1970; Guille and Grisvard, 1971; Schilperoort *et
al.*, 1973; Zaenen *et al.*, 1974). However, there are indications suggesting that direct nucleic acid interaction may not necessarily be
involved in causing plant tumours (Kostoff, 1930; Gautheret, 1948;
Braun, 1965): the plant tumour could thus be very different from that

* Agricultural Research Council Unit of Developmental Botany,
Cambridge.

found in animals. The search for the tumour inducing principle in plants has been difficult and conflicting results have been obtained. No definite hypothesis can yet be made about the mechanism of tumour induction.

We have therefore again investigated the possible relationship between the DNA of tumourous and non-tumorous tissues from sunflower stem and the DNA of *Agrobacterium tumefaciens*, by hybridisation of the plant DNA with complementary RNA transcribed from the bacterial DNA. Our aim has been to determine the extent of any such hybridisation and to find whether the homology is so exact as to allow us to say that Agrobacterium DNA can be found in the tumour tissue, or whether gross mismatching between the two partners is apparent.

There are many ways of designing such an experimental programme which have been tried in other systems. The methods of choice depend as much on the form that the foreign nucleic acid takes in the host cell as on the availability of different techniques and the ease with which they can be used in this system. In our experiments we have chosen to use an RNA copy of the Agrobacterium DNA: this is partly because of the ready availability of techniques for hybridisation of RNA to DNA and also because it is relatively easy to prepare radioactive RNA of very high specific activity by transcription of Agrobacterium DNA with *E. coli* polymerase *in vitro*.

Two types of hybridisation experiments might be done with such complementary RNA (cRNA). In one, the DNA of the plant tissue is saturated with increasing amounts of cRNA. The amount of hybrid formed under saturating conditions would then give a measure of the percentage of the plant DNA which contains sequences complementary to the Agrobacterium cRNA. This however, gives no information about the proportion of the cRNA which is complementary to the plant DNA; it may be that only a very small fraction of the Agrobacterium genome is homologous with or incorporated into the plant tumour cell. The method also suffers from the effects of traces of impurities in the cRNA, since high concentrations of cRNA are used. Small amounts of mismatched hybrids would lead to a situation where no defined saturation plateau would be obtained. On the other hand, this is the most direct method for detecting whether any Agrobacterium sequences can be found in plant DNA and the method is sensitive.

The alternative approach would be to use a vast excess of DNA and measure the kinetics of its reassociation from the denatured state using trace amounts of Agrobacterium cRNA as probe. Under these conditions, the proportion of the cRNA which hybridises would give a measure of the proportion of the bacterial genome which is homologous to the plant genome. However the method requires large amounts of DNA in very pure condition, since only a very small fraction of this DNA will ever hybridise to the radioactive RNA under these conditions.

We have chosen for our experiments the first and most direct of these methods, coupled with an analysis of the stability of the hybrids formed. The plant DNA was denatured and bound to millipore filters. Hybridisation to Agrobacterium cRNA was carried out by standard techniques (Birnstiel *et al.*, 1968; Bishop, 1969). The hybrids formed were melted by heating at increasing temperatures to determine their stability. The amount of plant DNA used and the specific radioactivity of the cRNA were such that it should be possible to detect one genome equivalent of Agrobacterium per genome of plant DNA.

RESULTS

AGROBACTERIUM COMPLEMENTARY RNA.

Complementary RNA was synthesised from Agrobacterium DNA using *E. coli* polymerase, according to the method described by Burgess (1969). The specific activity of the cRNA varied from 10^4 to 10^6 cpm/µg, depending on the preparation.

The cRNA showed a certain degree of self-complementarity (up to 15% after incubation in 2 x SSC at $65°$ for 16 hrs, at a concentration of 5.5 µg/ml), indicating that to some extent the polymerase transcribed the Agrobacterium DNA symmetrically. It hybridised specifically to Agrobacterium DNA and not at all to Proteus DNA. At concentrations up to about 10 µg/ml, it hybridised to not more than 4% of the Agrobacterium DNA: however, a reciprocal plot suggested that about 30% of the Agrobacterium DNA had been transcribed. Thus it is unlikely that the whole of the Agrobacterium DNA is represented in the cRNA.

SENSITIVITY OF THE HYBRIDISATION.

In order to estimate the smallest amount of Agrobacterium sequences which would be detectable, we hybridised Agrobacterium cRNA to decreasing amounts of Agrobacterium DNA, in the presence of mouse DNA as a heterologous carrier. The bacterial DNA was present in amounts equivalent to 1, 10 and 100 genomes of Agrobacterium per mouse genome. As shown in Table 1, the lowest input of Agrobacterium DNA, being

Table 1

Sensitivity of the hybridisation system.[*]

DNA		Percent of Agrobacterium DNA hybridised cRNA input (µg/ml)		Percent of total DNA on filter hybridised (x 10^4) cRNA input (µg/ml)	
		4	9.2	4	9.2
Mouse	23.4µg	-	-	18	-
" + Agrobacterium,	2.68µg	1.88	3.20	1,985	3,380
" + Agrobacterium,	0.268µg	2.23	3.44	270	450
" + Agrobacterium,	0.0268µg	1.27	3.70	33	74
	Average	1.79	3.44		
Sunflower callus	20 µg	-	-	57	132
Sunflower tumour	20 µg	-	-	52	127

Estimated number of Agrobacterium genome equivalents in plant DNA.

Sunflower callus				3.18	3.81
Sunflower tumour				3.07	3.79

[*] DNA was denatured and loaded on Millipore filters as described by Birnstiel *et al*. (1968). Filters were washed with 6 x SSC and dried under vacuum at 80° for 2 hrs. Hybridisation was carried out in 50% formamide, 6 x SSC at 55° for 40 hrs at two concentrations of cRNA as indicated. Specific activity of ^{3}H-cRNA was 8 x 10^5 cpm/µg). Filters were washed extensively in 6 x SSC at room temperature, then for 5 mins at 60° in 2 x SSC and finally incubated for 30 mins in 2 x SSC with 20 µg/ml pancreatic ribonuclease at room temperature. Radioactivity was measured by scintillation counting in a toluene-based scintillator.

1/1,000 of the amount of mouse DNA, was readily detectable. Also, the percentage of the added Agrobacterium DNA hybridised to the cRNA was independent of the amount present on the filter. The apparent number of Agrobacterium genomes present in plant DNA samples can thus be obtained from the percentage of plant DNA hybridised and the percent-

age of Agrobacterium DNA hybridised in a control sample run under the
same conditions.

HYBRIDISATION TO PLANT DNA.

An attempt was made to estimate the apparent number of Agrobacterium genomes in the DNA from callus and tumour cells. "Reconstructed"
samples were obtained by mixing small amounts of Agrobacterium DNA
(equivalent to 10 and 100 Agrobacterium genomes) with sunflower leaf
DNA, as with the mouse DNA described above. Tables 1 and 2 show
examples of the results obtained. It is clear that tumour and callus
DNA show very similar levels of hybridisation. Samples with the
addition of Agrobacterium DNA give values for the percentage of
hybridisation of cRNA to the bacterial DNA, from which we conclude
that both tumour and callus DNA apparently contain sequences homologous to Agrobacterium DNA in an amount which is equivalent to about
3 genomes of Agrobacterium per genome of plant. Estimates of between
1 and 10 genome equivalents were obtained, in the sunflower system,
when hybridisation was performed either in 2 x SSC at 65°C for 16 hrs
or in 50% formamide at 55° for 40 hrs (Birnstiel *et al*., 1972) (Table
3). The latter condition should favour the reaction of sequences of
a high GC content such as that of Agrobacterium DNA (58% GC) and reduce the rate of hybridisation of sequences with a low proportion of
GC such as that of plant DNA (Birnstiel *et al*., 1972). Thus, hybridisation in the presence of formamide at the temperature optimum for
the particular content of GC of the nucleic acid under investigation
should provide conditions which are more stringent than those in
2 x SSC alone (McConaughy *et al*., 1969).

Similar extents of hybridisation were obtained with most samples of
plant DNA, whether from callus or from tumour tissue cultures. Some
examples with other plant systems are shown in Table 4. The DNA from
sunflower leaf usually gave a lower level of hybridisation than the
tumour and callus DNA, a finding already mentioned by Schilperoort
et al. (1973) for the tobacco system.

Since the callus tissue used as control in these experiments had
been obtained from the same sunflower plant as the tumour which was
induced by Agrobacterium, we considered the possibility that Agrobacterium DNA had been taken up by much of the plant but only the
region of injection of the bacterium grew into a tumour. Indeed, in
some experiments DNA obtained from a sunflower callus from a different plant, which had not been injected with Agrobacterium, did show

Table 2

Hybridisation of Agrobacterium DNA to plant DNA.[*]

DNA	Total DNA on filters (µg)	Percent DNA hybridised (x 10^4) cRNA input (µg/ml)		
		2.3	3.6	5.5
Leaf	14.7 µg 14.70	6.3	16.5	20.4
" + Agrobacterium	2.5 µg 17.20	480	1250	1690
" + Agrobacterium	0.25 µg 14.95	106	186	273
Sunflower callus	11.30 µg	19.3	36.2	81.5
Sunflower tumour	10.6 µg	15.3	63.5	70.7

		2.3	3.6	5.5
Percent of DNA that would hybridise under these experimental conditions, if 1 genome equivalent of Agrobacterium was present in it:		4.53	9.12	13
Estimated number of genome equivalents in Sunflower callus		4.25	4.00	6.30
Sunflower tumour		3.40	7.00	5.45

[*]Hybridisation was carried out in 2 x SSC, 1% SDS at 69° for 17 hrs. Specific activity of ^{3}H-cRNA was 5.5 x 10^4 cpm/µg. Filters were washed for 1 hr in 2 x SSC at 69° (Bishop, 1969) and treated with 20 µg/ml RNAase for 30 mins at room temperature.

a lower level of hybridisation; however, we have not been able to reproduce this result, (Callus B, Table 4).

We also considered the possibility that the apparent hybridisation which we obtained was not the formation of a true RNA-DNA hybrid, but was an artifact due to impurities present in the DNA preparations. Plant DNA in general is notoriously difficult to purify from various contaminants such as polysaccharides and the extracts from tumour tissues were always more contaminated than those from callus. This is not the case with mammalian DNA, as shown by the low background level of counts bound to mouse DNA, though we have observed a higher level of noise with calf spleen DNA.

The results in Table 4 show that the extent of purification of the DNA had little effect on the extent of the hybridisation obtained, and it is unlikely therefore that the hybridisation observed is spurious and due to impurities.

Table 3
Hybridisation of Agrobacterium cRNA
to sunflower DNA under different conditions

% DNA Hybridised (x 10^4) and (No. of genome equivalents)		cRNA µg/ml	Conditions of hybridisation	Filter washing procedure
Sunflower Callus	Sunflower Tumour			
25 (2.9)	27 (3.2)	2.8	2xSSC, 1% SDS,	1 hr 65°C, 2xSSC.
91 (4.0)	102 (4.5)	10.6	18 hrs, 65°C.	
50 (5.9)	47 (5.6)	8	2xSSC, 1% SDS, 18 hrs, 65°C.	3 hrs 6xSSC r.t., 5' 65°C, 2xSSC.
42 (7.5)	49 (8.7)	4.7	50% Formamide, 6xSSC, 40 hrs, 55°C.	3 hrs 6xSSC r.t., 5' 50°C, 1xSSC.
28 (5.1)	29 (5.3)	2.6	2xSSC, 1% SDS,	1 hr 65°C, 2xSSC.
33 (3.3)	34 (3.4)	3.9	17 hrs 65°C.	
49 (4.7)	42 (4.0)	5.2		

It is clear, however, that there is some rather loose binding of RNA which behaves as though it were in a hybrid but which is either spurious or grossly mismatched. Such RNA can be removed from the filters by washing in 2 x SSC at 50°-60° after the ribonuclease treatment. Up to half the counts initially bound to the filters were removed in this way. This procedure was used in all the experiments presented above: the results thus show the apparent levels of hybridisation of RNA which is relatively firmly bound. The need for this high temperature wash is however an indication that the plant DNA behaves differently from Agrobacterium DNA: in the homologous hybridisation between Agrobacterium DNA and its own cRNA there is very little of such loosely bound RNA and the wash does not much reduce the number of counts bound to the filters. There is therefore a high level of identifiable noise due to the plant DNA. Any truly homologous Agrobacterium sequences in the plant DNA would however be expected to be firmly bound and unaffected by any extensive high temperature wash. The question then arises whether there is a series of partially homologous sequences which can hybridise to Agrobacterium cRNA, resulting in the production of mismatched hybrids. A proportion of such hybrids, according to the extent of mismatching, would be melted by

Table 4

Hybridisation of Agrobacterium cRNA
to various plant DNA.[*]

	cRNA input μg/ml	% DNA hybridised x 10^4
Sunflower callus A	)	50-91
Sunflower callus B	) 8-10 (	
from a non-infected plant	) (	19-21
Sunflower tumour C	) (	47
Sunflower tumour D	) 8 (	58
DNA further purified	) (	
Tobacco callus	) 8 (	37
tumour	) (	38
Carrot callus	) 5 (	56
tumour	) (	36

[*]The callus A and tumour C and D tissues were obtained from the
same plant; callus B was from a separate plant which had not
been exposed to Agrobacterium. The DNA of the tumour D was further purified by removal of some polysaccharide in alkali and
filtration of the denatured DNA through Whatman GF/A filters.

washing at increasing temperatures. In other words, there may be a
continuous spectrum of hybrids of different stabilities ranging from
"noise" to true homology. Some more exact criterion is therefore
needed to define homology. This problem was investigated in more detail by analysing the melting profiles of the hybrids formed.

MELTING PROFILES OF HOMOLOGOUS AND HETEROLOGOUS HYBRIDS.

The filters obtained after the hybridisation and counting were
washed free of scintillation liquid. Batches of duplicate filters
were incubated for five minutes in 1 x SSC preheated at the desired
temperatures. Figure 1 shows a comparison between the melting curves
obtained with the homologous hybrid and with the plant DNA-Agrobacter-
ium cRNA hybrid. It is apparent that very few counts are lost from
the filters in the case of the homologous hybrid up to about 60° and
that the hybrid then melts with a sharp melting profile with a Tm of
about 85°. On the other hand, with the plant DNA hybrids, about 25%
of the counts were lost at 60°. One could regard this treatment up
to 60° in 1 x SSC as a further washing procedure which differs from
the previous wash only in that it takes place at a lower salt concen-

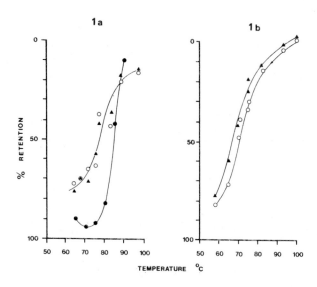

Fig. 1. Melting profiles of Agrobacterium cRNA-Agrobacterium DNA and Agrobacterium cRNA-sunflower DNA hybrids.

Melting and washing of the filter in warm 2 x SSC was carried out as described in the text, and in Table 1.

a) ●————● Agrobacterium DNA; o————o Sunflower tumour;
   ▲————▲ Sunflower callus.

b) o————o Sunflower tumour; ▲————▲ Sunflower callus.

Percent retention is calculated on the counts retained after washing at 50° in 1 x SSC, in addition to the usual washing procedure as above.

tration. The RNA lost under these conditions is either noise or very poorly matched hybrid, and of no significance in the present experiments. Above 60° the hybrids with both tumour and callus DNA gave similar melting profiles, with Tm in the region of 73° to 79°. Thus it seems that there may be a few sequences in any of the plant tissues which are truly homologous to Agrobacterium, but most of the hybrids are poorly matched. Some measure of the limits of detection is important to give significance to these results. Assuming that, from the calculations mentioned earlier, the whole melting profile of the hybrid with the plant DNA in this experiment represents the equival-

278

ent of approximately 3 to 4 Agrobacterium genomes per plant genome, then if only about 1/5 of it overlaps with the melting profile of the homologous hybrid, we can say that there is less than half of an Agrobacterium genome equivalent per plant genome.

We carried out similar experiments with tobacco control and tumour calluses and obtained essentially similar results. The melting curves of hybrids with tobacco DNA, shown in Figure 2 have a rather lower Tm of about 68°. This figure also shows that similar results were obtained with cRNA transcribed from a DNA satellite induced in Agrobacterium by mitomycin (W. Schuch, in preparation).

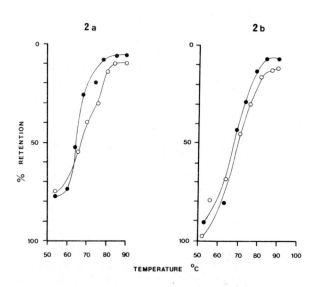

Fig. 2. Melting profiles of Agrobacterium cRNA-tobacco DNA hybrids.
Agrobacterium cRNA was transcribed from Agrobacterium DNA extracted from bacteria which had been incubated for 1½ hrs in the presence of 2 µg/ml Mitomycin C. DNA was banded in neutral CsCl and the main band DNA (p = 1.723 g cm⁻³) and light satellite DNA (p = 1.718 g cm⁻³) were used as templates in the *in vitro* system. Percent retention is calculated on the counts retained after washing at 50° in 1 x SSC.

a) melting profile with cRNA of Agrobacterium main band DNA.
b) melting profile with cRNA of Agrobacterium "satellite" DNA.

o——o Tobacco callus DNA; ●——● Tobacco tumour DNA.

We conclude that both normal and tumour plant DNAs show some hybridisation to Agrobacterium cRNA suggesting some degree of homology. The lowered Tm, however, shows that this homology is not perfect, but involves quite a high degree of mismatching between the nucleotide sequences. Since 1.5% mismatching causes a lowering of melting temperature of 1° (Bautz and Bautz, 1964), our results suggest that there are at least 10 to 15% of mismatched bases in the hybrids formed.

It remains a possibility that the plant tumour contains a portion of the Agrobacterium which is too small to be seen by these techniques. For example, the integration of a small phage DNA would be near the limits of detection. It is also conceivable that any foreign DNA in the tumour cell, whether integrated or not, is specifically lost during extraction. We consider this to be most unlikely, since several methods of extraction of DNA were used.

Therefore the simple idea that the plant tumour contains one or more genomes of Agrobacterium DNA is incorrect, and one must look further for the tumour inducing principle. It may be, that there is some significance in the finding that all the plant tissues examined contain some DNA which is partially homologous to Agrobacterium DNA.

ACKNOWLEDGEMENT

This work was supported by a grant from the Cancer Research Campaign.

REFERENCES

Bautz, E.K.F., and Bautz, F.A. (1964). The influence of noncomplementary bases on the stability of ordered polynucleotides. *Proceedings of the National Academy of Sciences, U.S.A.* 52, 1476-1481.

Birnstiel, M., Speirs, J., Purdom, I.F., Jones, K., and Loening, U.E. (1968). Properties and composition of the isolated ribosomal DNA satellite of *Xenopus laevis. Nature.* 219, 454-463.

Birnstiel, M.L., Sells, B.H., and Purdom, I.F. (1972). Kinetic complexity of RNA molecules. *Journal of Molecular Biology.* 63, 21-39.

Bishop, J.O. (1969). The effect of genetic complexity on the time-course of ribonucleic acid - deoxyribonucleic acid hybridisation. *Biochemical Journal.* 113, 805-811.

Braun, A.C. (1965). The reversal of tumour growth. *Scientific American.* 213, 75-83.

Burgess, R.R. (1969). A new method for the large scale purification of *Escherichia coli* deoxyribonucleic acid-dependent ribonucleic acid polymerase. *Journal of Biological Chemistry.* 244, 6160-6167.

Dulbecco, R. (1969). Cell transformation by viruses. *Science.* 166, 962-968.

Gautheret, R.J. (1948). Patologie végétale sur la culture de trois types de tissu de Scorsonère; tissu normaux tissu de crown gall et tissu accoutumés a l'hètèroauxin. *Compte Rendue Hebdomadaire des Séances de l'Academie des Sciences,* Paris. 226, 270-271.

Green, M. (1970). Oncogenic viruses. *Annual Review of Biochemistry.* 39, 701-756.

Guille, E., and Grisvard, J. (1971). Liberation of *Agrobacterium tumefaciens* DNA from the crown gall tumour cell DNA by shearing. *Biochemical and Biophysical Research Communications.* 44, 1402-1409.

Kostoff, D. (1930). Tumours and other malformations in certain Nicotiana hybrids. *Zentralblatt für Bacteriologie Parasitenkunde, Infektionskrankheiten und Hygiene.* Abt. II. 81, 244-260.

McConaughy, B.L., Laird, C.D., and McCarthy, B.J. (1969). Nucleic acid reassociation in formamide. *Biochemistry.* 8, 3289-3295.

Schilperoort, R.A., Sittert van N.J., and Schell, J. (1973). The presence of both Phage PS8 and *Agrobacterium tumefaciens* A_6 DNA base sequences in A_6 induced sterile crown gall tissue cultures *in vitro. European Journal of Biochemistry.* 33, 1-7.

Spiegelman, S., Burny, A., Das, M.R., Keydar, J., Schlom, J., Travnicek, M., and Watson, K. (1970). Characterization of the products of RNA-directed DNA polymerases in oncogenic RNA viruses. *Nature.* 227, 563-567.

Srivastava, S., and Chadha, K.C. (1970). Polymerases in oncogenic RNA viruses. *Biochemical and Biophysical Research Communications.* 40, 968-972.

Temin, H.M., and Mizutani, S. (1970). RNA-dependent DNA polymerase in virions of Rous Sarcoma virus. *Nature.* 226, 1211-1213.

Zaenen, I., Larebeke, van N., Teuchy, H., Montagu, van M., and Schell, J. (1974). Supercoiled circular DNA in crown gall-inducing Agrobacterium strains. *Journal of Molecular Biology.* 86, 109-127.

THE USE OF MUTANTS OF *AGROBACTERIUM TUMEFACIENS*
FOR INVESTIGATING THE MECHANISM OF CROWN GALL
TUMOUR INDUCTION IN PLANTS

D. COOPER

Genetics Department,
John Innes Institute, Colney Lane,
Norwich NR4 7UH.

The nature of the crown gall phenomenon is outlined.
Since crown gall tumours can be induced only by living
cells of *Agrobacterium tumefaciens*, it is suggested
that non-pathogenic mutants of this organism offer useful ways to study the disease. Methods of isolating
such mutants are discussed.

The mechanism of induction of crown gall tumours in plants is still
largely unknown; however, several well-established features of the
phenomenon make it of great interest. The role of the causative bacterium is apparently transient since cells of *Agrobacterium tumefaciens* need to be in contact with damaged plant tissue for only about
1-4 days to cause a tumour. Thereafter, living bacteria may be eliminated by heat treatment and the transformed tissue continues to grow
(White, 1945; Braun, 1943; Braun and Stonier, 1958). Explants of
the tumour will then grow indefinitely on plant tissue culture media
without growth substances (or hormones) of the auxin and cytokinin
type. It is growth in the absence of plant hormones which distingui-
shes crown gall tissue from normal plant tissue, but not from hormone
independent tissue which may arise in other ways, for example habituated tissue, or tissue from the spontaneous tumours of certain plant
hybrids. (These were discussed in detail by Butcher, 1973).

There is little evidence of host specificity for *A. tumefaciens*
among dicotyledonous plants; virulent strains will induce tumours on
plants of widely differing families and genera. No monocotyledons
have been found susceptible to crown gall disease. *Agrobacterium* is
thus specific for some aspect of dicotyledon physiology.

Research on crown gall broadly divides into two main areas: (1)
the search for the hypothetical chemical trigger by which *A. tumefaciens* induces tumours (designated "tumour inducing principle" by Braun
and Mandle, 1948; (2) the study of chemical and physiological differences between normal plant tissue and bacterium-free tumour tissue in

culture (reviewed by Meins, 1972). The difficulty in identifying the chemical trigger (or triggers) is the inability to induce tumours in plants except by placing viable cells of *A. tumefaciens* in contact with plant tissue. Extracts of *Agrobacterium* or of its spent growth medium, or bacteria killed by heat, U.V. or antibiotics are ineffective. While there are preliminary reports of gall production by RNA or DNA preparations from *A. tumefaciens* (Swain and Rier, 1972; Kado *et al.*, 1972), in no case has tissue culture evidence for the induction by isolated nucleic acids of hormone-independent tissue been presented. Tumour induction by DNA from PS8 phage, reported by Leff and Beardsley (1970) could not be repeated by other workers (Beiderbeck *et al.*, 1973). For this reason, the mechanism of tumour induction has been studied by rather indirect means.

Two parameters whose influence on tumour induction were first investigated are temperature and time. In many plants, tumour induction does not occur at temperatures above $30^\circ C$ which will allow growth of the bacteria and of tumour tissue induced at a lower temperature (Braun, 1947). Also plant cells are found to be susceptible to transformation for only a short time after tissue wounding. This is the "susceptible period", and correlates with the time of repair cell division induced by wounding (Braun, 1952; Lipetz, 1966). Introduction into the plant of viable *A. tumefaciens* cells during the susceptible period results in continued cell division, giving rise to a tumour. Induction may be viewed as a suppression of the mechanism switching off wound repair divisions (Braun, 1962). Beardsley (1972) gives a full account of this aspect.

Very few bacteria are needed to induce a tumour; it has been shown that infection with 1-10 viable cells of *A. tumefaciens* may be sufficient (Hildebrand, 1942). I have also observed that inocula containing, on average, about one viable cell of *A. tumefaciens* generally produce galls in sunflower stems.

A very precise way of perturbing biological systems is by mutation, and non-pathogenic mutants of *A. tumefaciens* are potentially of value in analysing the pathogen-host relationship in crown-gall. The observation that the loss of a chemical component of *A. tumefaciens*, resulting from a mutation, leads to loss of pathogenicity would be good evidence that the component is involved in tumour induction. Some of these components might be indirectly involved, such as those required for maintainance of viability of bacteria after inoculation into the plant. Such mutants could be identified in a preliminary screening.

Plant mutants of altered susceptibility to crown gall bacteria would be of great interest, but difficult to obtain. A definitive test of the hypothesis that incorporation and expression of *A. tumefaciens* information in plant cells occurs in crown gall induction (Guillé and Quétier, 1970) would be to detect the mutant gene product (for example, a temperature-sensitive enzyme) synthesized by crown gall tissue induced by a mutant strain of *A. tumefaciens*. This has not been achieved. In an analogous way, virus mutants have been valuable in studies of transformation of animal cells (Eckhart, 1972). In this article, I shall discuss observations made with different strains of *A. tumefaciens*.

Ways in which non-pathogenic mutants of *A. tumefaciens* have been obtained are summarised in Table 1. Spontaneous non-pathogenic mutants occur very rarely in *A. tumefaciens* populations, so nitrosoguanidine (NTG) mutagenesis has often been used, followed by random screening of survivors in test plants, usually *Helianthus annuus*. Langley and Kado (1972), and Davis and Rothman (1973), using optimum conditions for NTG mutagenesis, obtained high frequencies of non-pathogenic mutants (Table 1). Some of the auxotrophs isolated, particularly those requiring histidine, leucine or tryptophan, were found to be nonpathogenic in sunflower (Langley and Kado, 1972). Since NTG is prone to cause multiple mutations (Guerola *et al.*, 1971), one cannot conclude that auxotrophy and loss of pathogenicity arise from the same mutation unless simultaneous reversion of both characters can be demonstrated. Davis and Rothman concluded that since similar frequencies of auxotrophic and non-pathogenic mutants were obtained, similar numbers of genetic loci may control both characters.

Using sub-optimal conditions for NTG mutagenesis, I obtained a much lower frequency of non-pathogenic mutants than of auxotrophs. Three non-pathogenic mutants were obtained from 1,850 clones tested on plants of *Bryophyllum tubiflorum*. A similar frequency was observed in *Helianthus annuus*. Several adenine auxotrophs isolated after NTG mutagenesis of strain B6 gave slowly growing tumours in sunflower plants, but produced tumour callus of normal growth rate from *Nicotiana tabacum* (White Burley).

The pathogenicity of auxotrophic mutants was studied in detail by Lippincott and Lippincott (1966) and Schmidt *et al.* (1969) who used the frequency of tumour induction on Pinto bean leaves to measure the infectivity of strains of *A. tumefaciens* (Lippincott and Heberlein, 1965). Adenine, methionine and asparagine auxotrophs were found to

Table 1

Some non-pathogenic mutants of

Agrobacterium tumefaciens

Parent strain	Mutant phenotype selected	Induction	Other characters identified	Reference
A6	glycine resistance	spontaneous	initiates tumours when supplemented with auxin	Hendrickson *et al.*, (1934); Braun and Laskaris (1942)
B6	glycine resistance	spontaneous	-	Beardsley (1972)
B6	thioproline resistance	spontaneous	loss of one surface antigen	Hochster and Cole (1967)
B6	5-fluoro-tryptophan resistance	spontaneous and NTG mutagenesis	-	D. Cooper (unpublished)
various	resistance to bacter-iocin produced by *A. tumefaciens* strain 84	spontaneous	-	Kerr and Htay (1974)
IAM1525	abnormal cell morphology	NTG mutagenesis	-	Fujiwara and Fukui (1972)
C-58 Ach	rapid growth at 36°C	growth at 36°C	nutritional re-quirements altered. Not specifically identified	Hamilton and Fall (1971)

have low infectivity values. Prototrophic revertant strains, and mutant strains supplied with their growth requirements after infection, were found to have infectivities nearer wild-type values. This suggests that, to induce tumours, A. tumefaciens must retain the capability of continued protein and nucleic acid synthesis after inoculation into the plant. In line with this idea, Beiderbeck (1970) has shown that rifampicin, an antibiotic that inhibits DNA-dependent RNA synthesis, blocks tumour induction by sensitive A. tumefaciens. A mutant strain of the bacterium resistant to rifampicin induced tumours even in the presence of the antibiotic, showing that the latter was acting specifically on the bacteria, not the plant cells. In view of the apparently transient role of A. tumefaciens in tumour induction, the requirement for macromolecular synthesis during this process is of interest. It could be investigated further by the use of A. tumefaciens mutants temperature-sensitive for DNA, RNA or protein synthesis.

In addition to random screening, non-pathogenic mutants have been obtained by selection methods. The earliest of these was repeated sub-culture of wild-type strains through medium supplemented with sub-inhibitory concentrations of glycine. This procedure reproducibly yields non-pathogenic strains (Longley et al., 1937; Van Lanen et al., 1952). Glycine can inhibit cell-wall synthesis and lead to spheroplast formation (Rubio-Huertos and Beltrá, 1962) but the relationship of glycine resistance to loss of pathogenicity in A. tumefaciens is not clear. The non-pathogenic strain A66, obtained by glycine treatment of wild-type strain A6, was found to give tumours only if the inoculation site in the plant was supplied with auxin (indole butyric acid). Auxin alone, or cells of A66 alone, caused only small swellings, but together, tumours similar to those produced by the pathogenic parent, were induced (Braun and Laskaris, 1942).

Hochster and Cole (1967) found a non-pathogenic mutant of A. tumefaciens strain B6 by selecting a strain highly resistant to thioproline. The inhibitory action of thioproline in A. tumefaciens is not known, but in E. coli K-12 it is reported to inhibit the formation of prolyl t-RNA (Unger and De Moss, 1966). Hochster and Cole (1967) further found that the non-pathogenic mutant lacked one of four antigens possessed by the wild-type strain. It is of interest to compare this observation with the results of Graham (1971) who made serological comparisons of A. tumefaciens and A. radiobacter. Isolates of the latter are taxonomically placed with A. tumefaciens, but are non-pathogenic (De Ley *et al.*, 1972). Graham found that, whereas most *A. tumefaciens* strains possessed four antigens, all *A. radiobacter* isolates had only three. The variable antigen is not perfectly corre-lated with pathogenicity, since Graham found that a few pathogenic wild isolates of *A. tumefaciens* also lacked the fourth antigen. It would be of great value if an immuno-chemical change resulting from mutational loss of pathogenicity could be identified in *A. tumefaciens*.

I recently found that resistance to tryptophan analogues may be used to enrich non-pathogenic mutants among survivors of mutagenesis. Non-pathogenic mutants were found at a frequency of about 6% among mutants of strain B6 resistant to 150 μg/ml 5-fluorotryptophan (wild-type is inhibited by 45 μg/ml). A similar frequency was found among 5-fluorotryptophan resistant strains occurring spontaneously in a non-mutagenized culture. Non-pathogenic strains also occur frequently among mutants of B6 resistant to α-methyl tryptophan. The mechanisms of resistance to tryptophan analogues in *A. tumefaciens* are not worked out, but in *Pseudomonas putida* some 5-fluorotryptophan-resistant strains analysed show altered regulation of tryptophan biosynthetic enzymes (Maurer and Crawford, 1971). The non-pathogenic *A. tumefac-iens* mutants obtained in this way are still able to convert trypto-phan to indole acetic acid. The reason for the loss of pathogenicity is unknown.

Morphological mutants have been isolated from *A. tumefaciens* strain IAM 1525 by Fujiwara and Fukui (1972) by visual screening of cells after NTG mutagenesis. Twenty mutants were obtained, 14 of which were abnormal (spherical) at 37°C and normal at 27°C. Three formed branched cells at the higher temperature only and three were spherical at both temperatures. Two of these mutants were non-pathogenic, for unknown reasons.

Stonier (1960) first reported that *A. tumefaciens* strain T37 pro-duced a bacteriocin, but this has not been studied in detail. More recently, Kerr and Htay (1974) have examined many wild isolates of *Agrobacterium* for bacteriocin production. One strain of *A. radio-bacter* (non-pathogenic), number 84, produces a bacteriocin to which many Australian isolates of *A. tumefaciens* are sensitive. The bacter-iocin is rapidly diffusible and probably of low molecular weight. A most interesting observation was that bacteriocin-resistant mutants of the sensitive indicator strains were all found to have lost pathogeni-city. The resistant strains were isolated as spontaneous mutants growing in the inhibition zone of the indicator lawn (Roberts and Kerr,

1974). The mutants were confirmed to be *A. tumefaciens*, and were serologically and biochemically indistinguishable from their bacteriocin-
sensitive (pathogenic) parents. It was suggested that, in common with
many colicins, the bacteriocin of *Agrobacterium* operates *via* a surface attachment site on the sensitive cell, and that this surface
structure is lost on mutation to resistance.

Another interesting report of induction of non-pathogenic strains
is that of Hamilton and Fall (1971). They found a very high proportion of non-pathogenic clones by culturing *A. tumefaciens* strains C-58
or Ach at 36°C. Alterations in nutritional characteristics were
found in the non-pathogenic derivatives, and also in the pathogenic
clones, after culture at 36°C. These observations have not yet been
clarified.

Attachment of cells of *A. tumefaciens* to specific receptor sites
in plant tissue has been suggested by Lippincott and Lippincott (1969)
and Beaud *et al.* (1972). The former authors showed that the infectivity of strain B6 was reduced by mixing it with a culture of the non-
pathogenic strain IIBNV6. Quantitative estimates of infectivity were
obtained using the Pinto bean leaf assay (Lippincott and Heberlein,
1965). Inhibition of infectivity was not seen if leaves were inoculated with the non-pathogenic strain shortly after infection with B6.
Tumour induction by strain B6 was also suppressed by the non-pathogenic mutant Pa-82 (Beaud *et al.*, 1972). This occurred if inoculation
of pea seedlings with Pa-82 preceded B6. If the order of inoculation
was reversed, normal tumours appeared. Both observations were interpreted in terms of specific sites in the wounded plant tissue to which
A. tumefaciens cells may attach; the non-pathogenic mutants were
then assumed to block these sites irreversibly, preventing access by
pathogenic strains inoculated subsequently. There is little direct
evidence for the existence of such sites in plant tissue.

A property of *A. tumefaciens* which has received a lot of attention
is that of lysogeny. Since the report of Beardsley (1955) that strain
B6 harbours a prophage, omega, it has been speculated that a temperate
phage is closely involved in pathogenicity. By U.V. irradiation,
Beardsley (1955) produced a "cured" derivative of strain B6 (B6-806)
which is sensitive to phage omega. This strain, cured of its prophage,
was found by Beardsley to be as pathogenic in plants as the wild-type
strain. In addition, *A. tumefaciens* strain V-1, when cured of its
prophage LV-1 by incubation at 37°C, was found to be as pathogenic

as the parent strain (Brunner and Pootjes, 1969). No sequence homology between DNA of the cured strain V-1C and DNA from phages of the
omega group (i.e. omega, PS8, PB2A, LV-1) could be detected (De Ley
et al., 1972). These strains provide strong evidence that omega
group phages are not involved in tumour induction. It is of interest
that hybridization between RNA complementary to phage PS8 DNA and DNA
from crown gall tissue has been detected (Schilperoort *et al.*, 1973).
The nature of the hybrid is described by Schilperoort *et al.* (this
Symposium).

Transfer of virulence between strains of *Agrobacterium* has been
reported by Kerr (1971). Pathogenic donor strains were classified in
biotype 2 in the system of Keane *et al.* (1970). They do not grow on
Schroth's medium, and do not produce ketolactose. Non-pathogenic
recipients were of biotype 1, and could be selected on the medium of
Schroth. Some recipients were marked with chloramphenicol or novobio-
cin resistance genes and could be distinguished from donors on medium
containing these antibiotics. The pathogenic donors were inoculated
into tomato plants, then genetically marked non-pathogenic strains
were applied to the original inoculation sites. After a month, bacteria were re-isolated from galls, and clones of the recipient strains
selected. The transfer was very efficient, since 45%-75% of clones of
effective recipients had acquired virulence. Kerr reports that all
attempts to transfer virulence between strains by mixed culture in
medium were unsuccessful.

A system of recombination in *Agrobacterium* has not been developed.
A system for genetic analysis of non-pathogenic mutants would provide
useful information complementary to that obtained from biochemical
work. For example, after NTG mutagenesis it would be desirable to
test, by conjugation or some other system of genetic transfer, whether
a selected mutant phenotype, such as resistance to an amino acid analogue, arises from the same mutation as one causing loss of pathogenicity. This test is important since NTG is prone to give multiple
mutations (Guerola *et al.*, 1971).

Some experiments in this direction have been carried out. Kern
(1969) demonstrated transformation of *A. tumefaciens* to sulphonamide
resistance by DNA from *Rhizobium leguminosarum*. The identity of
transformants was checked serologically, and the transformation frequency was approximately 10^{-5}. Klein and Klein (1953) and Kern (1965)
also reported transfer of pathogenicity to non-pathogenic bacteria by
DNA preparations from *A. tumefaciens*.

Milani and Heberlein (1972) defined conditions for transfection, in which DNA extracted from phage LR-4 was taken up and expressed by *A. tumefaciens* strain B66, as shown by subsequent plaque formation. Recipient cells were competent for a period of 4 hrs in early exponential phase in broth culture. It was estimated that DNA equivalent to approximately 7.9×10^4 phage genomes must be supplied for each bacterium which is infected.

Plasmid transfer to *Agrobacterium* has been demonstrated. Cole and Elkan (1973) showed that an R factor carrying penicillin G resistance could be transmitted to *A. tumefaciens* from *Rhizobium japonicum* at frequencies of 0.25% or 0.1% of recipients, depending on the strain of *A. tumefaciens*. Co-transfer of resistance to penicillin G, neomycin and chloramphenicol was also observed. Datta *et al.* (1971) transferred the *Pseudomonas aeruginosa* R factor RP4, carrying resistances to penicillin, kanamycin and tetracycline, from an *E. coli* K-12 strain to *A. tumefaciens* with a frequency of 10^{-8}-10^{-9}. There are no reports, as yet, of chromosome transfer by conjugation or transduction in *Agrobacterium*.

Different facets of the crown gall phenomenon are shown by various wild-type isolates of *Agrobacterium*. For example, *A. radiobacter* referred to earlier is a group of isolates which are taxonomically close to *A. tumefaciens*, but are non-pathogenic in plants. *A. rhizogenes* (Riker *et al.*, 1928) induces massive proliferation of roots in susceptible species, and only a limited amount of undifferentiated callus.

Certain *A. tumefaciens* isolates, e.g. strain T37, induce disorganized aggregates of shoots and leaves called teratomata (Kersters *et al.*, 1973; Butcher, 1973).

A recent study by Beiderbeck (1973) showed that root production by *A. rhizogenes* on *Kalanchoë* leaves is suppressed by spraying infected plants with dilute solutions of kinetin; undifferentiated tumours were obtained instead. Root formation was also suppressed if leaves were co-infected with *A. rhizogenes* and the teratoma inducing *A. tumefaciens* strain T37. The nature of the growths induced by *Agrobacterium* may be governed by the auxin/cytokinin ratio produced at infection sites. It is known that *Agrobacterium* can produce substances of the auxin (Kaper and Veldstra, 1958) and cytokinin (Upper *et al.*, 1970) types, but the mechanism of regulation of the levels of these substances by bacteria infection is largely speculative at this stage.

Chemical components of crown gall tissues which have received

attention recently are the guanidines octopine, nopaline, lysopine
and arginine (Petit *et al.*, 1970, and references therein).   It was
found that different wild-types of *A. tumefaciens* would induce galls
containing high levels of either octopine or nopaline, and that the
guanidine observed is the same as that which the inducing strain of
bacteria can specifically degrade (Petit *et al.*, 1970).   The interpretation of these observations is not yet clear, however, since Seitz
and Hochster (1964) and Johnson *et al.* (1974) have observed the occurrence of unusual guanidines, in much lower amounts than in crown gall,
in normal plant tissue.   The specific function of the compounds octopine, lysopine and nopaline in plant cells is not known.

It must be concluded that, although much work has been done, the
crown gall problem is far from solved.   As long as it is not possible
to induce tumours with cell-free extracts of *A. tumefaciens*, the most
informative experiments will be those in which appropriate mutant
strains of bacteria are used as controls.   The use of mutants could
also be extended by studying, for example, complementation between
non-pathogenic strains.   In recent experiments, I have found that
mixed inoculation of sunflower plants with pairs of non-pathogenic
mutants may give rise to tumours.   Control plants, inoculated with
the mutants singly, did not develop tumours.   The nature of the interaction between the mutants is currently being investigated.

REFERENCES

Beardsley, R.E. (1955). Phage production by crown gall bacteria and
the formation of plant tumors. *American Naturalist*. 89, 175-176.

Beardsley, R.E. (1972). The inception phase in crown gall disease.
Progress in Experimental Tumor Research. 15, 1-75.

Beaud, R., Beardsley, R.E., and Kurkdjian, A. (1972). Bacteriophages
of *Agrobacterium tumefaciens* III.   Effect of attenuation of virulence on lysogeny and phage production. *Canadian Journal of
Microbiology*. 18, 1561-1567.

Beiderbeck, R. (1970). Untersuchungen an Crown gall IV. Rifampicin
und ein resistenter Klon von *Agrobacterium tumefaciens* bei der
Tumorinduktion. *Zeitschrift für Naturforschung*. 25b, 1458-1460.

Beiderbeck, R. (1973). Wurzelinduktion an Blättern von *Kalanchoë
daigremontiana* durch *Agrobacterium rhizogenes* und der Einfluss von
Kinetin auf diesen Prozess. *Zeitschrift für Pflanzenphysiologie*.
68, 460-467.

Beiderbeck, R., Heberlein, G.T., and Lippincott, J.A. (1973). On the
question of crown gall tumor initiation by DNA of bacteriophage

PS8. *Journal of Virology.* 11, 345-350.

Braun, A.C. (1943). Studies on tumor inception in the crown gall disease. *American Journal of Botany.* 30, 674-677.

Braun, A.C. (1947). Thermal studies on the factors responsible for tumor initiation in crown gall. *American Journal of Botany.* 34, 234-240.

Braun, A.C. (1952). Conditioning of the host cell as a factor in the transformation process in crown gall. *Growth.* 16, 65-74.

Braun, A.C. (1962). Tumor inception and development in the crown gall disease. *Annual Review of Plant Physiology.* 13, 533-558.

Braun, A.C. and Laskaris, T. (1942). Tumor formation by attenuated crown gall bacteria in the presence of growth promoting substances. *Proceedings of the National Academy of Sciences, U.S.A.* 28, 468-477.

Braun, A.C., and Mandle, R.J. (1948). Studies on the inactivation of the tumor inducing principle. *Growth.* 12, 255-269.

Braun, A.C. and Stonier, T. (1958). Morphology and physiology of plant tumors. *Protoplasmatologia.* 10, part 5a. Vienna, Springer-Verlag.

Brunner, M., and Pootjes, C.F. (1969). Bacteriophage release in a lysogenic strain of *Agrobacterium tumefaciens. Journal of Virology.* 3, 181-186.

Butcher, D.N. (1973). The origins, characteristics and culture of plant tumour cells. In *Plant Tissue and Cell Culture.* H.E. Street, Editor. Botanical monograph no. 11, 356-391. Blackwell.

Cole, M.A., and Elkan, G.H. (1973). Transmissible resistance to penicillin G, neomycin and chloramphenicol in *Rhizobium japonicum. Antimicrobial Agents and Chemotherapy.* 4, 248-253.

Datta, N., Hedges, R.W., Shaw, E.J., Sykes, R.B., and Richmond, M.H. (1971). Properties of an R-factor from *Pseudomonas aeruginosa. Journal of Bacteriology.* 108, 1244-1249.

Davis, C.H., and Rothman, R.H. (1973). Induction of crown gall by nitrosoguanidine-treated *Agrobacterium tumefaciens. Mutation Research.* 20, 283-285.

De Ley, J., Gillis, M., Pootjes, C.F., Kersters, K., Tytgat, R., and Van Braekel, M. (1972). Relationship among temperate *Agrobacterium* phage genomes and coat proteins. *Journal of General Virology.* 16, 199-214.

Eckhart, W. (1972). Oncogenic viruses. *Annual Review of Biochemistry.* 41, 503-516.

Fujiwara, T., and Fukui, S. (1972). Isolation of morphological mutants of *Agrobacterium tumefaciens*. *Journal of Bacteriology*. 110, 743-746.

Graham, P.H. (1971). Serological studies with *Agrobacterium radio-bacter*, *Agrobacterium tumefaciens* and *Rhizobium* strains. *Archiv für Mikrobiologie*. 78, 70-75.

Guerola, N., Ingraham, J.L. and Cerdá-Olmedo, E. (1971). Induction of closely linked multiple mutations by nitrosoguanidine. *Nature New Biology*. 230, 122-125.

Guillé, E., and Quétier, F. (1970). Le crown gall: modèle expérimental pour l'application du mécanisme de régulation quantitative de l'information génétique à l'évènement néoplasique. *Bulletin du Cancer*. 57, 217-238.

Hendrickson, A.A., Baldwin, I.L., and Riker, A.J. (1934). Studies on certain physiological characters of *Phytomonas tumefaciens*, *Phytomonas rhizogenes* and *Bacillus radiobacter*. *Journal of Bacteriology*. 28, 597-618.

Hamilton, R.H., and Fall, M.Z. (1971). The loss of tumour initiating ability in *Agrobacterium tumefaciens* by incubation at high temperature. *Experientia*. 27, 229-230.

Hildebrand, E.M. (1942). A micrurgical study of crown gall infection in tomato. *Journal of Agricultural Research*. 65, 45-59.

Hochster, R.M., and Cole, S.E. (1967). Serological comparisons between strains of *Agrobacterium tumefaciens*. *Canadian Journal of Microbiology*. 13, 569-572.

Johnson, R., Guderian, R.H., Eden, F., Chilton, M.-D., Gordon, M.P., and Nester, E.W. (1974). Detection and quantitation of octopine in normal plant tissue and crown gall tumors. *Proceedings of the National Academy of Sciences, U.S.A.* 71, 536-539.

Kado, C.I., Heskett, M.G., and Langley, R.A. (1972). Studies on *Agrobacterium tumefaciens*: characterization of strains 1D135 and B6, and analysis of the bacterial chromosome, transfer RNA and ribosomes for tumor inducing ability. *Physiological Plant Pathology*. 2, 47-57.

Kaper, J.M., and Veldstra, H. (1958). On the metabolism of tryptophan by *Agrobacterium tumefaciens*. *Biochimica et Biophysica Acta*. 30, 401-420.

Keane, P.J., Kerr, A., and New, P.B. (1970). Crown gall of stone fruit. II Identification and nomenclature of *Agrobacterium* isolates. *Australian Journal of Biological Sciences*. 23, 585-595.

Kern, H. (1965). Untersuchungen zur genetischen Transformation zwischen *Agrobacterium tumefaciens* und *Rhizobium* spez. I Ubertragung der

Fähigkeit für induktion pflanzlichen tumoren auf *Rhizobium* spez.
Archiv für Mikrobiologie. 51, 140-155.

Kern, H. (1969). Interspezifische Transformation zwischen *Agrobacterium tumefaciens* und *Rhizobium leguminosarum*. *Archiv für Mikrobiologie*. 66, 63-68.

Kerr, A. (1971). Acquisition of virulence by non-pathogenic isolates of *Agrobacterium radiobacter*. *Physiological Plant Pathology*. 1, 241-246.

Kerr, A., and Htay, K. (1974). Biological control of crown gall through bacteriocin production. *Physiological Plant Pathology*. 4, 37-44.

Kersters, K., De Ley, J., Sneath, P.H.A., and Sackin, A. (1973). Numerical taxonomic analysis of *Agrobacterium*. *Journal of General Microbiology*. 78, 227-239.

Klein, D.T., and Klein, R.M. (1953). Transmittance of tumor-inducing ability to avirulent crown gall and related bacteria. *Journal of Bacteriology*. 66, 220-228.

Langley, R.A., and Kado, C.I. (1972). Conditions for mutagenesis by N-methyl-N'-nitro-N-nitrosoguanidine and relationships of *A. tumefaciens* mutants to crown gall tumor induction. *Mutation Research*. 14, 277-286.

Leff, J., and Beardsley, R.E. (1970). Action tumorigène de l'acide nucléique d'un bactériophage présent dans les cultures de tissu tumoral de tournesol (*Helianthus annuus*). *Comptes Rendus de l'Académie des Sciences (Paris) Series D*. 270, 2505-2507.

Lipetz, J. (1966). Crown gall tumorigenesis II. Relations between wound healing and the tumorigenic response. *Cancer Research*. 27, 1597-1605.

Lippincott, J.A., and Heberlein, G.T. (1965). The quantitative determination of the infectivity of *Agrobacterium tumefaciens*. *Journal of Bacteriology*. 90, 1155-1156.

Lippincott, B.B., and Lippincott, J.A. (1966). Characteristics of *Agrobacterium tumefaciens* auxotrophic mutant infectivity. *Journal of Bacteriology*. 92, 937-945.

Lippincott, B.B., and Lippincott, J.A. (1969). Bacterial attachment to a specific wound site as an essential stage in tumor initiation by *Agrobacterium tumefaciens*. *Journal of Bacteriology*. 97, 620-628.

Longley, B.J., Berge, T.O., Van Lanen, J.M., and Baldwin, I.L. (1937). Changes in the infective ability of *Rhizobium* and *Phytomonas tumefaciens* induced by culturing on media containing glycine. *Journal of Bacteriology*. 33, 29.

Maurer, R., and Crawford, I.P. (1971). New regulatory mutation affect-

294

ing some of the tryptophan genes in *Pseudomonas putida*. *Journal of Bacteriology*. 106, 331-338.

Meins, F. (1972). Stability of the tumor phenotype in crown gall tumors of tobacco. *Progress in Experimental Tumor Research*. 15, 93-109.

Milani, V.J., and Heberlein, G.T. (1972). Transfection in *Agrobacterium tumefaciens*. *Journal of Virology*. 10, 17-22.

Petit, A., Delhaye, S., Tempé, J., and Morel, G. (1970). Recherches sur les guanidines des tissus de crown gall. Mise en évidence d'une relation biochimique spécifique entre les souches d'*Agrobacterium tumefaciens* et les tumeurs qu'elles induisent. *Physiologie Végétale*. 8, 205-213.

Riker, A.J., Banfield, W.M., Wright, W.H., and Keith, G.W. (1928). The relation of certain bacteria to the development of roots. *Science*. 68, 357-359.

Roberts, W.P., and Kerr, A. (1974). Crown gall induction: serological reactions and sensitivity to mitomycin-C and to bacteriocin of pathogenic and non-pathogenic strains of *Agrobacterium radiobacter*. *Physiological Plant Pathology*. 4, 81-91.

Rubio-Huertos, M., and Beltrá, R. (1962). Fixed pathogenic L-forms of *Agrobacterium tumefaciens*. *Nature*. 195, 101.

Schilperoort, R.A., Van Sittert, N.J., and Schell, J. (1973). The presence of both phage PS8 and *Agrobacterium tumefaciens* A6 DNA base sequences in A6 induced sterile crown gall tissue cultured *in vitro*. *European Journal of Biochemistry*. 33, 1-17.

Schmidt, R.M., Lippincott, B.B., and Lippincott, J.A. (1969). Growth requirements and infectivity of auxotrophic adenine-dependent mutants of *Agrobacterium tumefaciens*. *Phytopathology*. 59, 1451-1454.

Seitz, E.W., and Hochster, R.M. (1964). Lysopine in normal and crown gall tumor tissue of tomato and tobacco. *Canadian Journal of Botany*. 42, 999-1004.

Stonier, T. (1960). *Agrobacterium tumefaciens* Conn. II Production of an antibiotic substance. *Journal of Bacteriology*. 79, 889-898.

Swain, L.W., and Rier, J.P. (1972). Cellular transformation in plant tissue by RNA from *Agrobacterium tumefaciens*. *Botanical Gazette*. 133, 318-324.

Unger, L., and De Moss, R.D. (1966). Action of a proline analogue, L-thiazolidine-4-carboxylic acid, in *E. coli*. *Journal of Bacteriology*. 91, 1556-1563.

Upper, C.D., Helgeson, J.P., Kemp, J.D., and Schmidt, C.J. (1970).

Gas-liquid chromatographic isolation of cytokinins from natural sources. *Plant Physiology*. 45, 543-547.

Van Lanen, J.M., Riker, A.J., and Baldwin, I.L. (1952). The effect of amino acids and related compounds upon the growth, virulence and enzyme activity of crown gall bacteria. *Journal of Bacteriology*. 63, 723-734.

White, P.R. (1945). Metastatic (graft) tumors of bacteria-free crown galls on *Vinca rosea*. *American Journal of Botany*. 32, 237-241.

IS THERE FOREIGN DNA IN CROWN GALL TUMOUR DNA?

MARY-DELL CHILTON, S.K. FARRAND, FRANCINE EDEN,
T. CURRIER, A.J. BENDICH, M.P. GORDON AND E.W. NESTER.

*University of Washington, Seattle,
Washington 98195, U.S.A.*

We have analyzed crown gall tumour DNA for traces of
Agrobacterium tumefaciens DNA and PS8 bacteriophage DNA
by three different methods, including one method sensitive enough to detect less than one genome equivalent
per tumour cell, and in no case have we found any. In
addition, we have now looked in a preliminary way for
Agrobacterium plasmid DNA in tumour DNA and likewise we
have found none. From all of these studies taken together, we conclude that there is at present no convincing nucleic acid hybridization evidence for foreign DNA
integration as the molecular basis of crown gall tumourigenesis.

Fundamental questions about the molecular basis of crown gall
tumourigenesis are currently in dispute. A tumour, initiated by inoculation of a wounded plant with *A. tumefaciens*, can be propagated
in tissue culture in the absence of viable bacteria. Tumour tissue
in vitro maintains a transformed phenotype indefinitely on selective
medium: it grows without addition of hormones required by normal
callus tissue. It is of the utmost importance in understanding the
mechanism of neoplastic transformation in plants, and possibly in animal systems as well, to discover whether the transformed phenotype is
due to a permanent alteration of the nuclear DNA of the transformed
cell (genetic change) or to some less permanent form of alteration
(epigenetic change).

One type of experiment which supports an epigenetic model for crown
gall transformation is the observation that tumourous tissue can revert
to what appears to be normal plant tissue. Braun (1959) observed
that after cloned teratoma tissue was successively grafted to apices
of a series of healthy plants, normal fertile shoots were produced.
A leaf morphology marker assured that the normal shoots derived from
tumour rather than host tissue (Braun, in press). Sacristán and
Melchers (1969), working with undifferentiated tumour tissue which
was not cloned, demonstrated that on suitable medium shoots, roots,

leaves and flowers are produced. However, they pointed out the possibility that untransformed cells were present in the tumour culture. Recent attempts to achieve reversion of cloned unorganized tumour tissue have not yet succeeded (Melchers, 1971 and private communication). Thus the possibility of reversion remains in doubt in the case of fully transformed unorganized tumour tissue.

A number of different experimental approaches have provided evidence which seems to support a genetic change as the molecular basis of transformation in crown gall tumours. *A. tumefaciens* DNA (Schilperoort *et al.*, 1967, 1973; Quetier *et al.*, 1969; Srivastava, 1970; Milo and Srivastava, 1969) and PS8 bacteriophage DNA (Schilperoort *et al.*, 1973), *A. tumefaciens* RNA (Milo and Srivastava, 1969), bacterial antigens (Schilperoort *et al.*, 1969; Chadha and Srivastava, 1971; De Torok and Cornesky, 1971) and viable PS8 phage particles (Parsons and Beardsley, 1968; Kurkdjian, 1970; Tourneur and Morel, 1970) have all been reported in axenic crown gall tumour tissue. Viable phage were not detected by other investigators (Schilperoort *et al.*, 1973; Farrand, unpublished data). DNA isolated from nuclei of axenic tumour tissue was reported to hybridize with labelled RNA complementary to *A. tumefaciens* DNA (Schilperoort, 1967), suggesting integration of foreign DNA into the chromosomes of the host plant. Estimated levels of bacterial DNA in tumour DNA range from 0.2% (Quetier *et al.*, 1969) to 0.9% (Schilperoort *et al.*, 1973), and PS8 phage DNA is reported to constitute 1.8% (Schilperoort *et al.*, 1973) of tumour DNA. The latter finding is most remarkable because the inciting bacterial strain appeared by several criteria to be free from detectable PS8 bacteriophage, and PS8 DNA sequences could not be detected in the DNA from this bacterium by nucleic acid hybridization (Schilperoort *et al.*, 1973).

Unfortunately, most of the hybridization studies of foreign nucleic acids in crown gall tumours have failed to include the Tm, the mean thermal dissociation temperature for the duplexes detected. This measurement is essential in establishing the identity of the duplexes as authentic bacterial base sequences. Thermal dissociation profiles in one system (Schilperoort *et al.*, 1973) have been reported separately (Van Sittert, 1972), and revealed that control "duplexes" formed with normal callus DNA had Tm values only 2° lower than those formed with tumour DNA, while the latter were 8° below homologous duplex Tm values. Thus the nucleic acid hybridization evidence for foreign DNA in crown gall tumours is not unequivocal.

In a different type of hybridization study, we have failed to detect any bacterial or PS8 DNA in crown gall tumour DNA under conditions allowing detection of 0.01% bacterial and 0.001% PS8 DNA (Chilton et al., 1974). The basis of these experiments is the measurement of the renaturation kinetics of labelled probe (bacterial or PS8 phage) DNA before and after addition of large excesses of tumour or control DNA. The putative bacterial DNA in the tumour DNA should increase the concentration, and therefore the rate of renaturation, of the probe DNA. Reconstruction experiments with mixtures of salmon DNA plus authentic bacterial or phage DNA have demonstrated the sensitivity of this technique (Chilton et al., 1974). These results ruled out the presence of as much as one bacterial genome per three diploid tumour cells. However, they were not sufficiently sensitive to rule out one phage genome per diploid tumour cell.

We now report a more sensitive assay for the presence of PS8 bacteriophage DNA sequences in three tumour lines which have been described previously (Schilperoort et al., 1973; Chilton et al., 1974; Johnson et al., 1974): B6-806(PS8) Xanthi tobacco tumour induced by a lysogen constructed in this laboratory; B6Sch tobacco tumour, also a Seattle line; and A6Sch, a White Burley tobacco tumour isolated at the University of Leiden and kindly provided by Dr. R.A. Schilperoort. Figure 1 shows the results of an attempt to detect PS8 DNA in these three tumour lines. The advantage of plotting our renaturation data as a function of P_0t, the product of probe DNA concentration and time, has been discussed (Chilton et al., 1974). The increased specific activity of the probe DNA in this experiment (14×10^6 cpm/µg, compared to 5.4×10^5 cpm/µg in our earlier one (Chilton et al., 1974)) allows an increase in the sensitivity of detection, such that approximately one phage genome per three diploid tumour cells could be detected. The reconstruction experiments show that as little as 0.0042 µg/ml PS8 DNA produces a recognizable shift in the renaturation kinetics of probe DNA in this experiment (Fig 1). The diploid tobacco nucleus contains 6×10^{12} Daltons of DNA (Johnson et al., 1974), and one PS8 phage particle, 4.1×10^7 Daltons (Chilton et al., 1974). Thus, one phage genome per diploid tumour cell should constitute 6.8×10^{-4}%. The data of Figure 1 show that 1.77-2.1 mg/ml tumour DNA contains less than 0.0042 µg/ml of PS8 DNA, i.e., the presence of 2×10^{-4}% (2 parts per million or one phage genome per three diploid tumour cells) is ruled out.

300

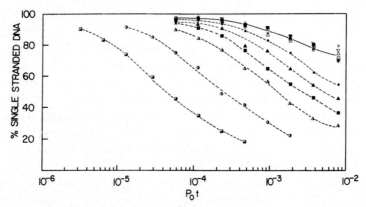

Fig 1. Renaturation kinetics of ^{32}P-labelled bacteriophage DNA probe in the presence of tumour DNA, control DNA and model mixtures.

^{32}P-labelled PS8 bacteriophage DNA, isolated as described (Chilton *et al.*, 1974), sheared by French pressure cell (*ca.* 10^6 Daltons), was adjusted to 0.004 µg/ml in 2 M NaClO$_4$, 0.015 M NaH$_2$PO$_4$, 0.015 M Na$_2$HPO$_4$. Sheared unlabelled DNA, isolated as described (Chilton *et al.*, 1974), was dialyzed against distilled water, evaporated to dryness, redissolved in the labelled DNA solution, and samples (10 µl) were sealed in capillaries. Renaturation kinetic measurements were performed as described elsewhere (Chilton *et al.*, 1974) using the hydroxylapatite assay for duplex formation. Renaturation temperature was 25° below Tm of the probe DNA. Unlabelled DNAs added were:

o A6Sch tumour, 2.1 mg/ml
∇ B6-806(PS8) tumour, 1.86 mg/ml
▼ B6Sch tumour, 1.77 mg/ml
△ Normal tobacco callus, 1.59 mg/ml
□ Salmon, 2.15 mg/ml
● 0.0042 µg/ml PS8 + 2.0 mg/ml salmon
▲ 0.009 µg/ml PS8 + 2.15 mg/ml salmon
■ 0.018 µg/ml PS8 + 2.15 mg/ml salmon
▲ 0.042 µg/ml PS8 + 2.03 mg/ml salmon
◑ 0.175 µg/ml PS8 + 2.09 mg/ml salmon
◪ 0.93 µg/ml PS8 + 2.23 mg/ml salmon
x No added DNA

The tumour tissues studied were unfortunately not cloned, and could contain a mixture of transformed and untransformed cells. The calculation of the sensitivity of the hybridization experiment rests on the assumption that most of the tumour callus consists of transformed cells. In support of this view, we have observed that unorganized tumour tissue does not "feed" normal callus tissue *in vitro*. Normal callus co-cultivated with unorganized tumour tissue on hormone-free agar (with a Millipore membrane divider between the two) grows extremely poorly compared with the tumour callus. Teratomata gave a strikingly different result, and in fact, fed normal callus so successfully that the normal callus outgrew the teratoma tissue. Only unorganized tumours have been used in this study; we believe it unlikely that these tumour tissues contain a significant fraction of normal (untransformed) cells.

Another significant reservation must be pointed out about the hybridization technique employed here and earlier (Chilton *et al.*, 1974). Although the renaturation kinetic approach is extremely sensitive for detecting copies of the whole probe genome, as our reconstruction experiments demonstrate, a small fraction of the genome ($\approx$ 5%) present in single or even multiple copies, would escape detection. As we have argued elsewhere (Eden *et al.*, 1974), multiple copies of a small region of bacterial or phage genomes might be detectable by hybridization studies employing filter-bound tumour DNA and either labelled DNA or complementary RNA synthesized from template DNA of *A. tumefaciens* or PS8 phage.

In order to assess the sensitivity of filter-bound DNA reactions for detecting foreign DNA in tumour DNA, we have performed reconstruction experiments using model mixtures of phage or bacterial DNA with salmon DNA, immobilized on filters. In a systematic survey, we attempted to detect the decreasing levels of bacterial or phage DNA in such model filters using labelled cRNA (Eden *et al.*, 1974) and labelled DNA (Farrand *et al.*, submitted). The model filter experiments show that cRNA hybridization can detect 0.01% phage DNA or 1% bacterial DNA; however tumour DNA, after rigorous purification, fails to bind significant amounts of either type of cRNA (Eden *et al.*, 1974). The DNA/DNA-filter hybridization system proved to be at least 10-fold less sensitive; technical problems restricted the sensitivity for tumour DNA filters even further, and no high-melting duplexes were observed between tumour DNA and labelled phage or bacterial DNA (Farrand *et al.*, in preparation). Thus the hybridization data purporting to demonstrate bacterial (Schilperoort *et al.*, 1967; Quetier *et al.*, 1969; Srivastava, 1970; Schilperoort *et al.*, 1973) and phage (Schilperoort *et al.*, 1973) DNA in crown gall tumour DNA were not confirmed in our laboratory despite a careful exploration of the two hybridization techniques with model systems. We concluded that no convincing hybridization evidence was available supporting the model of foreign DNA integration as the molecular basis of crown gall tumourigenesis (Chilton *et al.*, 1974; Eden *et al.*, 1974; Farrand *et al.*, submitted)

Although renaturation kinetic experiments described above and elsewhere (Chilton *et al.*, 1974) rule out the presence of the entire phage or bacterial genome, the possible presence of an *A. tumefaciens* plasmid was not ruled out. Evidence suggesting that plasmid DNA may be involved in the process of tumourigenesis has already been reported by Zaenen *et al.* (1974). These workers observed a large plasmid DNA (90-140 x 10^6 Daltons) in 11 virulent natural isolates of *A. tumefaciens* which was absent in 8 avirulent strains. Data from our laboratory strongly support the conclusion that the plasmid has an important role in tumourigenesis. We have found that the conversion of *A. tumefaciens* strain C58 from virulence to avirulence by incubation at elevated temperature, first reported by Hamilton and Fall (1971), is accompanied by loss of a large plasmid detectable in the virulent C58 strain (Watson *et al.*, in preparation). We have also found, as first observed by Kerr and his collaborators with other *A. tumefaciens* strains (Roberts and Kerr, 1974; Kerr and Htay, 1974), that spontaneous mutants of strain C58 which are resistant to bacteriocin 84 are avirulent. In addition, we found that such mutants lack the large plasmid which is carried by the parent C58 strain (Watson *et al.*, in preparation). Our findings thus substantiate the suggestion of Zaenen *et al.* (1974) that the plasmid is essential to the expression of virulence.

Although plasmid-borne genetic information could be involved at any of several steps in the process of tumour induction, we have explored the possibility that plasmid DNA sequences persist in the transformed plant cell. Since the plasmid, with genome size of 10^8 Daltons, represents approximately 5% of the total bacterial genome size (2 x 10^9 Daltons) (Chilton *et al.*, 1974), the presence of the entire plasmid could have escaped detection in our earlier renaturation kinetic experiments (Chilton *et al.*, 1974). The presence of a small fraction of the plasmid sequences most assuredly would have escaped detection, since it would represent a negligible percentage of the labelled total bacterial DNA. Accordingly, we have repeated our earlier investigation, this time using highly labelled plasmid DNA as a probe to look for plasmid DNA sequences in tumour DNA.

The presence of intact, closed-circular plasmid DNA in a lysate of
A. tumefaciens B6-806 prepared by a penicillin-sarkosyl-pronase method
is demonstrated by the appearance of radioactivity at a density greater than that of the bulk chromosomal DNA in a CsCl-ethidium-bromide
gradient (Fig 2) and the presence of a fast-sedimenting peak in alkaline sucrose gradients (Fig 3). Electron microscopic examination of
the pooled fractions from the CsCl-ethidium bromide gradient peak confirms the presence of a large circular plasmid DNA (Plate 1). From
contour length measurements, we estimate the molecular weight of this
plasmid to be 105 x 10^6 ± 6 x 10^6.

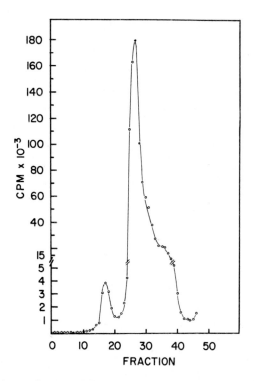

Fig 2 Isolation of Plasmid DNA by CsCl-Ethidium Bromide Density Gradient Centrifugation.
A 1 ml sample of a penicillin-Sarkosyl-pronase lysate from a culture of *A. tumefaciens* B6-806 (labelled with 50 μCi/ml ^{3}H-thymi-
dine, New England Nuclear, 40 Ci/mM) was sheared on a vortex mixer at top speed for 30 secs and mixed with 5.2 g CsCl plus 3.9 ml
TES buffer (Zaenen *et al.*, 1974). To the solution was added 0.1 ml
of 5 mg/ml ethidium bromide in TES. The CsCl solution (n=1.3925)
was centrifuged to equilibrium in a Beckman ultracentrifuge (40
rotor, 3 days, 35,000 rpm, 15.5°C). A portion of each fraction
(5 μl) was used for assay of TCA-insoluble radioactivity.

304

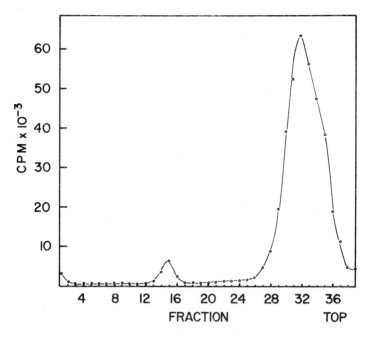

Fig 3. Detection of Plasmid DNA by Alkaline Sucrose Gradient Sedimentation Velocity.

A 0.1 ml portion of the lysate described in Figure 2 was layered
on a 5-20% linear sucrose gradient (0.3 M NaOH, 0.5 M NaCl, 0.02
M Versene) and centrifuged for 15 mins at 35000 rpm (18.5°C) in
a Beckman SW-39 rotor. Fractions were collected from the bottom
of the tube and assayed for TCA insoluble radioactivity.

The kinetic complexity of the plasmid DNA can be predicted, assuming that it contains no repeated sequences. We have observed the
$C_o t_{1/2}$ (Britten and Kohne, 1968) of sheared PS8 bacteriophage DNA in
0.15 M phosphate to be 4.4×10^{-2} (unpublished data), and the genome
size of this phage, using T7 phage DNA as standard, is therefore
4.1×10^7 (data not shown). The plasmid DNA, with a genome size of
1.05×10^8, is therefore expected to exhibit a $C_o t_{1/2}$ value, under
these conditions, of $(1.05 \times 10^8 / 4.1 \times 10^7) \times 4.4 \times 10^{-2} = 0.11$. Because
we isolated only minute amounts of plasmid DNA, the specific activity
of the plasmid cannot be measured directly. It can, however, be inferred. Using the observed renaturation kinetics and fitting them to
the expected $C_o t_{1/2}$ value calculated above, C_o for plasmid DNA can be
calculated, and specific activity inferred. By this technique (Fig 4)
the plasmid of *A. tumefaciens* B6-806 isolated from the dense peak in

Figure 2 has a specific activity of 1.48×10^6 cpm/μg. This value is similar to the specific activity of DNA in the main band of the gradient (Fig 2), estimated from its A_{260} (data not shown).

Renaturation kinetic analysis can also be used to determine whether bacterial DNA isolated by usual procedures includes plasmid DNA sequences. Total bacterial DNA was isolated from *A. tumefaciens* B6-806 after lysis with 1% sodium dodecyl sulphate (15', 70°) and pronase digestion, using Marmur's procedure (Marmur, 1961). Although this procedure does not allow the plasmid to be isolated as covalently closed circles, the plasmid is present in fragmented form, as Figure 4 shows. The addition of total bacterial DNA (172 μg/ml) decreases the $P_o t_{1/2}$ (Chilton *et al.*, 1974) of renaturation of 0.156 μg/ml labelled plasmid DNA 39-fold. Therefore, 172 μg of bacterial DNA must contain $38 \times 0.156 = 5.92$ μg of plasmid DNA. Thus, 3.45% of the total bacterial DNA is plasmid DNA.

In a separate study, using another lot of plasmid DNA isolated from a gradient similar to that of Figure 2, we have observed that the kinetics of reassociation of labelled B6-806 plasmid DNA are not influenced by the addition of a large excess of PS8 bacteriophage DNA (data not shown). This plasmid, therefore, has no base sequence homology with PS8 DNA.

Using the highly labelled plasmid DNA pooled from the gradient of Figure 2, we have attempted to detect plasmid DNA sequences in the DNA of crown gall tumour line B6-806, a Seattle line which has been previously described (Johnson *et al.*, 1974). Figure 5 shows that the kinetics of reassociation of 0.022 μg/ml labelled plasmid DNA in the presence of tumour DNA, normal callus DNA and salmon DNA are very similar. The reconstruction experiment containing salmon DNA plus 0.027 μg/ml plasmid DNA differs very significantly from these, and allows us to infer that 2.32 mg/ml of tumour DNA cannot contain as much as 0.027 μg/ml plasmi DNA. Thus 1.16×10^{-3}% plasmid DNA is ruled out. One plasmid genome (1.05×10^8 Daltons) per diploid tobacco cell (6×10^{12} Daltons) (Siegel *et al.*, 1973) would be 1.75×10^{-3}%. Thus our experiment rules out the presence of one plasmid copy per 1.5 diploid tumour cells. The technical reservations already mentioned, i.e. that the tumour line is not cloned and that a small fraction of the plasmid genome could escape detection, apply to the interpretation of this experiment also. Some additional uncertainties enter into the calculation of sensitivity in this case. Both the specific activity of the plasmid DNA and the amount of plasmid DNA in the re-

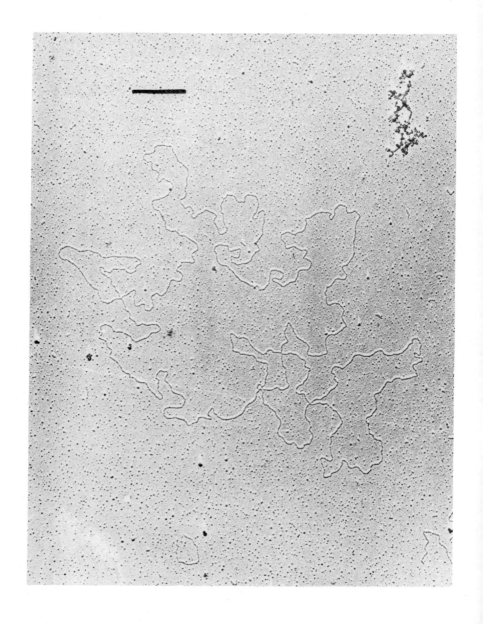

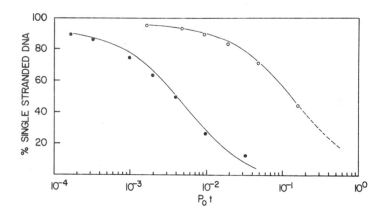

Fig 4. Renaturation kinetics of plasmid DNA.

The plasmid peak from the CsCl gradient of Figure 2 was pooled, extracted with isopropyl alcohol to remove dye, dialyzed, sheared and its renaturation kinetics measured as described (Chilton et al., 1974). The concentration was inferred from $t_{1/2}$ of the reaction, as described in the text. Labelled plasmid DNA (0.156 µg/ml) was found to renature more rapidly in the presence of total A. tumefaciens B6-806 bacterial DNA.

       o Plasmid DNA alone

       ● Plasmid DNA + B6-806 bacterial DNA (172 µg/ml)

Plate 1

Electron Micrograph of A. tumefaciens B6-806 Plasmid DNA.

Plasmid DNA from the dense peak of the CsCl gradient of Figure 2 was pooled, extracted with isopropyl alcohol to remove dye, and dialyzed. After standing at 5° for several days to allow relaxation of the circular DNA molecules, the DNA was spread on parlodian coated grids by the technique of Kleinschmidt (1968).   Bar = 1 µ.

308

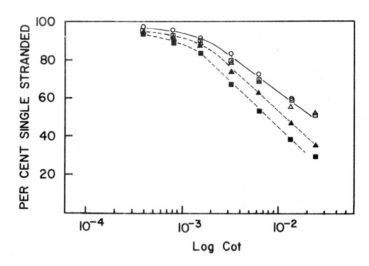

Fig 5 Renaturation kinetics of ^{3}H-labelled plasmid DNA probe in the presence of tumour DNA, control DNA and model mixtures.

The sheared plasmid DNA (0.022 µg/ml) described in Figure 4 was used as labelled probe in experiment similar to that of Figure 1. Renaturation conditions were as described in Figure 1. Reconstruction mixtures were prepared by adding B6-806 bacterial DNA to salmon DNA and assuming (see text and Figure 4) that 3.45% of the bacterial DNA is plasmid DNA. The P_0t plot has been described (Chilton et al., 1974). P_0 = probe DNA concentration in moles of nucleotide per liter. t = time in seconds. Duplex assay was performed by hydroxylapatite chromatography (Chilton et al., 1974). Unlabelled DNAs added were:

    ○ Salmon DNA (3.5 mg/ml)

    □ Normal tobacco callus DNA (1.66 mg/ml)

    △ B6-806 tumour DNA (2.32 mg/ml)

    ▲ Salmon DNA (3.6 mg/ml) + 0.027 µg/ml plasmid DNA

    ■ Salmon DNA (3.1 mg/ml) + 0.055 µg/ml plasmid DNA

---

construction experiments are inferred from renaturation data, which are subject to experimental error. The sensitivity of this experiment is thus less certain than in other experiments of this type. Our provisional conclusion is that there does not appear to be one plasmid copy per diploid tumour cell. Further experiments with greater sensitivity are planned.

REFERENCES

Braun, A.C. (1959). A demonstration of the recovery of the crown gall
    tumor cell with the use of complex tumors of single cell origin.
    Proceedings of the National Academy of Sciences, U.S.A. 45, 932-938.
Braun, A.C. (1974). "The Cell Cycle and Tumorigenesis" in The Cell
    Cycle and Differentiation, Springer (in press).
Britten, R.J., and Kohne, D.E. (1968). Repeated sequences in DNA.
    Science. 161, 529-540.
Chadha, K.C., and Srivastava, B.I.S. (1971). Evidence for the presence
    of bacteria-specific proteins in sterile crown gall tumor tissue.
    Plant Physiology. 48, 125-129.
Chilton, M.-D., Currier, T.C., Farrand, S.K., Bendich, A.J., Gordon,
    M.P., and Nester, E.W. (1974). *Agrobacterium tumefaciens* DNA and
    PS8 bacteriophage DNA not detected in crown gall tumors. *Proceed-
    ings of the National Academy of Sciences, U.S.A.* 71, 3672-3676.
De Torok, D., and Cornesky, R.A. (1971). Antigenic and immunological
    determinants of oncogenesis in plants infected with *Agrobacterium
    tumefaciens. Colloques Internationaux du Centre National de la
    Recherche Scientifique.* No. 193, 443-451.
Eden, F.C., Farrand, S.K., Powell, J., Bendich, A.J., Gordon, M.P.,
    and Nester, E.W. (1974). Attempts to detect DNA from *Agrobacterium
    tumefaciens* and bacteriophage PS8 in crown gall tumors by comple-
    mentary RNA/DNA-filter hybridization. *Journal of Bacteriology.*
    119, 547-553.
Farrand, S.K. (unpublished data).
Farrand, S.K., Eden, F.C., and Chilton, M.D. (Submitted for publica-
    tion, 1974).
Hamilton, R.H., and Fall, M.Z. (1971). The loss of tumour-initiating
    ability in *Agrobacterium tumefaciens* by incubation at high temper-
    ature. *Experentia.* 27(2), 229-230.
Johnson, R., Guderian, R.H., Eden, F.C., Chilton, M.-D., Gordon, M.P.,
    and Nester, E.W. (1974). Detection and quantitation of octopine in
    normal plant tissue and in crown gall tumors. *Proceedings of the
    National Academy of Sciences, U.S.A.* 71, 536-539.
Kerr, A., and Htay, K. (1974). The mechanism of biological control of
    crown gall. *Physiological Plant Pathology.* 4, 37.
Kleinschmidt, A.K. (1968). Monolayer techniques in electron microscopy
    of nucleic acid molecules. *Methods in Enzymology XII-B.* 361-377.
Kurkdjian, A. (1970). Observations sur la présence des phages dans les
    plaies infectees par differentes souches d'*agrobacterium tumefac-
    iens* (Smith et Town) Conn. *Annales de l'Institut Pasteur.* 118, 690-696.

310

Marmur, J. (1961). A procedure for the isolation of deoxyribonucleic
acid from micro-organisms. *Journal of Molecular Biology*. 3, 208-218.

Melchers, G. (1971). Transformation of habituation to autotrophy and
tumour growth and recovery. *Colloques Internationaux du Centre
National de la Recherche Scientifique*. No. 193. 229-234.

Milo, G.E., and Srivastava, B.I.S. (1969). RNA-DNA hybridization
studies with the crown gall bacteria and the tobacco tumor tissue.
Biochemical and Biophysical Research Communications. 34, 196-199.

Parsons, C.L., and Beardsley, R.E. (1968). Bacteriophage activity in
homogenates of crown gall tissue. *Journal of Virology*. 2, 651.

Quetier, F., Huguet, T., and Guillé, E. (1969). Induction of crown
gall: partial homology between tumor-cell DNA, bacterial DNA and
the G+C-rich DNA of stressed normal cells. *Biochemical and Biophysical Research Communications*. 34, 128-133.

Roberts, W.P., and Kerr, A. (1974). Crown gall induction: a study of
pathogenic and non-pathogenic strains of *Agrobacterium radiobacter*.
Physiological Plant Pathology. 4, 81-91.

Sacristán, M.D., and Melchers, G. (1969). The caryological analysis
of plants regenerated from tumourous and other callus cultures of
tobacco. *Molecular General Genetics*. 105, 317-333.

Schilperoort, R.A. (1967). Ph.D. Thesis, University of Leiden.

Schilperoort, R.A., Veldstra, H., Warnaar, S.O., Mulder, G., and Cohen,
J.A. (1967). Formation of complexes between DNA isolated from
tobacco crown gall tumours and RNA complementary to *Agrobacterium
tumefaciens* DNA. *Biochimica et Biophysica Acta*. 145, 523-525.

Schilperoort, R.A., Meys, W.H., Pippel G.M.W., and Veldstra, H.
(1969). *Agrobacterium tumefaciens* cross-reaction antigens in
sterile crown gall tumours. *FEBS Letters*. 3, 173-176.

Schilperoort, R.A., van Sittert, N.J., and Schell, J. (1973). The
presence of both phage PS8 and *Agrobacterium tumefaciens* A_6 DNA
base sequences in A_6-induced sterile crown gall tissue cultured
in vitro. *European Journal of Biochemistry*. 33, 1-7.

Siegel, A., Lightfoot, D., Ward, O.G., and Keener, S. (1973). DNA
complementary to ribosomal RNA: relation between genomic proportion and ploidy. *Science*. 179, 682-683.

Srivastava, B.I.S. (1970). DNA-DNA hybridization studies between
bacterial DNA, crown gall tumor cell DNA and the normal cell DNA.
Life Science. 9, part II, 889-892.

Tourneur, J., and Morel, G. (1970). Sur la présence de phages dans
les tissus de crown gall cultivés *in vitro*. *Compte rendue*

hebdomadaire des séances de l'Academie des sciences, Paris. 270, 2810-2812.

Tourneur, J., Garnier-Expert, D., and Morel, G. (1973). Origins des phages présents dans les tissus de crown gall cultivés *in vitro.* *Compte rendue hebdomadaire des séances de l'Academie des sciences, Paris.* 276, 2265-2267.

Van Sittert, N. (1972). Ph.D. Thesis, University of Leiden.

Watson, B., Nester, E.W., and Gordon, M.P. (Manuscript in preparation).

Zaenen, I., Van Larebeke, N., Teuchy, H., and Schell, J. (1974). Supercoiled circular DNA in crown gall inducing *Agrobacterium* strains. *Journal of Molecular Biology.* 86, 109-127.

WHO DOES WHAT IN THE LEGUME ROOT-NODULE?

J.E. BERINGER AND A.W.B. JOHNSTON

Department of Genetics, John Innes Institute,
Colney Lane, Norwich NR4 7UH, England.

Nodule formation and eventual nitrogen-fixation require
a continual interaction between the host plant and infecting Rhizobium. Genetic information from both partners must be expressed for this interaction to succeed.
The relative contribution of each symbiont in the regulation of this process is unknown. Nitrogenase and the
haeme component of leghaemoglobin are apparently coded
by bacterial genes, while the leghaemoglobin apoprotein
is of plant origin. Nitrogen fixation in the legume
nodule requires leghaemoglobin and possibly the induction
of the nitrogenase genes in the bacterium by the plant.
However, observations of nitrogenase activity in tissue
cultures and in the non-legume *Trema cannabina* suggest
that this activity is not dependent upon a complex legume-
Rhizobium interaction.

Other contributions in this section on plant-microbe interactions
have dealt with various aspects of the interaction between *Agrobacterium tumefaciens* and dicotyledonous plants. We shall discuss some
features of what appears to be a much more complex association between
the closely related bacterium Rhizobium (Graham, 1964; Gibbins and
Gregory, 1972) and various leguminous plants. We shall not review the
whole topic of legume symbiosis, as this has been done by a number of
others (Nutman, 1955, 1969; Dixon, 1969; Dart, 1974), but instead we
shall discuss some aspects of nodulation that may help an understanding of the interaction. The early stages of nodulation, associated
with strain recognition, root penetration and the initiation of cell
division, are essential prerequisites for the formation of the functional nodule. The exclusion of these topics from discussion reflects
only a space limitation and not a diminished view of their importance
(see review by Dart, 1974).

The title perhaps implies a clear division of labour between the
legume and Rhizobium in their contributions to the physiological and
structural changes involved in nodule formation and eventual nitrogen-
fixation, and a knowledge of what each partner is doing. This is not

the case. Knowledge of the system is very limited, due mainly, as will be discussed in more detail later, to an inability to modify it in any controlled manner and to the difficulty of studying a problem combining plant and bacterial gene interaction with developmental biology.

In contrast to the Agrobacterium-induced crown gall, cell proliferation in the legume nodule is organized and appears to be under strict genetic control. This is witnessed by the fact that the size, time of appearance, gross morphology and mean number of nodules per plant show great variation from strain to strain within a legume species; moreover, plant mutants defective in each of these characteristics have been described (Nutman, 1955, 1969). Apart from this endogenous control the nature of the infective Rhizobium strain also affects these characteristics. As a general rule, more effective bacterial strains produce fewer, larger nodules (Nutman, 1955; Dart, 1974). Thus, there is some feedback resulting from the presence of large effective nodules which inhibits further nodule initiation. Inhibitors of nodulation are known in legumes (Elkan, 1961; Phillips, 1971). One, produced by a mutant of soybean which fails to be nodulated (Williams and Lynch, 1954; Tanner and Anderson, 1963), is able to prevent nodulation of other legumes grown nearby, apparently without inhibiting the growth of the rhizobia (Elkan, 1961). This mutant may synthesize nodule inhibitor constitutively, but this has not been reported.

Nodule formation involves cell differentiation as well as cell proliferation, for example when a fully developed vascular system differentiates to serve the transport needs of the nodule. Can nodule formation, therefore be interpreted as a modification of developmental processes leading to the formation of normal organs, such as lateral roots? This could imply the absence of a whole battery of legume genes specifically concerned with nodulation. Rather, genes involved in other developmental processes might be concerned in nodulation, under the direction of a specific control system. This would be consistent with the occurrence of nodulation in non-legume, non-Rhizobium, nitrogen fixing symbioses (Becking, 1970) and in a non-legume (*Trema cannabina*) by a Rhizobium species (Trinick, 1973).

One of the most startling changes associated with the formation of an effective legume nodule is the presence of leghaemoglobin. The function of this oxygen-binding protein in the nodule has not been entirely resolved, though it has been shown to stimulate oxygen uptake

in isolated bacteroids. This increase is associated with an ele-
vation of nitrogenase activity (Wittenberg et al., 1974).

Leghaemoglobin is one of the few products of the symbiosis whose
genetic origin has been demonstrated. It is clear that at least the
apoprotein is determined by the legume. Properties of the apoprotein,
such as electrophoretic and chromatographic mobility, are independent
of the class of infecting Rhizobium (Dilworth, 1969; Cutting and
Schulman, 1971; Broughton and Dilworth, 1971). Conversely, when a
single Rhizobium strain is used to infect different legumes with dis-
tinguishable leghaemoglobins, these plant-specific differences in the
molecule are retained (Dilworth, 1969; Cutting and Schulman, 1971).
The recent finding (Verna, Nash and Schulman, 1974) that polysome-
associated mRNA isolated from soybean root nodules can act as a template for synthesis of leghaemoglobin in the cell-free wheat embryo
protein synthesis system demonstrated that no rhizobial component is
required for the synthesis of the apoprotein. Reasonable evidence
exists that the haeme group of the molecule is synthesized by the
bacteroids and is then incorporated into the leghaemoglobin apoprotein
(by the plant?) (Cutting and Schulman, 1969).

With a few exceptions, which will be mentioned later, the fixation
of nitrogen is correlated with the presence of leghaemoglobin in the
nodule. Let us consider two general explanations of the fact that
leghaemoglobin is not produced in nodules containing ineffective
strains of Rhizobium. (a) The ineffective strain fails to induce or
support leghaemoglobin synthesis for one reason or another; failure
of leghaemoglobin synthesis in turn prevents the expression of the
nitrogen-fixation genes. (b) The ineffective strain is defective in
some stage of nitrogenase synthesis and the expression or, at least,
the presence of the functional enzyme is the inducer of leghaemoglobin
synthesis.

We favour the second alternative since nitrogen-fixation by Rhizobium strains can certainly occur in the absence of leghaemoglobin, as
in the Trema cannabina nodule (M.J. Trinick, personal communication).
Furthermore, the only example of legume nodules containing leghaemoglobin but being, at first sight ineffective, occurs in a particular
mutant of the pea (Holl and La Rue, personal communication). These
nodules contain large polysaccharide particles and it has been proposed that these act as a physical barrier to carbohydrate transport,
as a result of which the bacteroids are starved of sources of energy.
If the nodules are cut open and provided with pyruvate or succinate,

nitrogenase activity, as measured by acetylene reduction, is detectable immediately, indicating that nitrogenase is present in these conditionally ineffective nodules.

The term ineffective has been used earlier without definition. An ineffective strain is one which can induce nodules, but which cannot fix nitrogen. It is perhaps misleading to classify all such mutants of Rhizobium under this single heading. In a study of one general class of ineffective strains (antimetabolite-resistant) by Pankhurst (1974) five different types of defect led to ineffectiveness. These ranged from the formation of tumour-like growths on the roots, devoid of Rhizobium, vascular system or meristem, to the case where the only defect was the occurrence of larger, more pleomorphic bacteroids than in the wild-type. Six of the mutants which failed to induce formation of infection threads within the nodule had altered cell envelopes which allowed leakage of internal antigens. This suggests that the bacterial cell envelope is important in determining the success of nodulation, a conclusion supported by the finding that all viomycin-resistant Rhizobium mutants are ineffective (Schwinghamer, 1964; Hendry and Jordan, 1969). Such resistance is conferred by accumulation of phospholipid and change of the cation-exchange capacity in the cell envelope (MacKenzie and Jordan, 1970; Yu and Jordan, 1971).

Resistance to other antibiotics (Schwinghamer, 1964), bacteriophages (Kleczkowska, 1950) or glycine (Staphorst and Strijdom, 1971) is sometimes associated with loss of effectiveness. However, it is not clear why some of these mutants should be rendered ineffective while others retain effectiveness. The problem might well be resolved if the target for the mutation leading to resistance was identified. For all these classes of mutants the basis of resistance might be an alteration in the structure of the cell envelope, which could effect permeability. In most cases this hypothesis has not been tested and, in any case, it is not clear how such an alteration would be reflected in a loss of effectiveness. Signals might be prevented from entering or leaving the bacterium, or perhaps the structure of the bacterial membrane itself has to be recognized by the plant.

Mutants isolated on the basis of their resistance to various agents are one of the classical mutational types in bacterial genetics. The other main group is of auxotrophs, isolated on account of nutritional deficiency. Studies using auxotrophy to examine the nature of the determinants of a successful symbiosis have revealed a variety of ineffective strains. A leucine auxotroph of R. *meliloti* in which rhizobia fail to leave the infection thread has been described (Dénarié, Truchet and Bergeron, 1975). A different leucine-requiring mutant of R. *meliloti* was also found to be associated with loss of effectiveness. In transduction experiments with this mutant, all prototrophic recombinants regained the ability to fix nitrogen (Kowalski and Dénarié, 1972), suggesting a causal relationship between leucine auxotrophy and loss of effectiveness. Some, but by no means all, purine auxotrophs are also ineffective (Schwinghamer, 1969; Scherrer and Dénarié, 1971). Dénarié *et al.* (1975) reported that reversion of ineffective ade⁻, leu⁻, ura⁻ and ilv⁻ auxotrophs restored effectiveness, suggesting that the growth requirement itself was responsible for ineffectiveness. However, a note of caution may be interpolated. A second mutation independent of the auxotroph, and the real cause of the loss of effectiveness, could be reverted along with the auxotrophy by extra-cistronic suppression.

A *pur*⁻ strain isolated by Schwinghamer (1969) failed to induce nodules on one variety of pea (Home Freezer), while on another variety (Freezonian) ineffective nodules were formed. This is a clear example of genotype-genotype interaction in the formation of a successful symbiosis. This variation of symbiotic ability was also observed with a riboflavin-requiring mutant of R. *trifolii*, where the level of effectiveness varied between clover varieties (Schwinghamer, 1970). With this particular mutant effectiveness was restored by the addition of riboflavine to the plant, while in its absence the rhizobia failed to develop into bacteroids (Pankhurst, Schwinghamer and Bergersen, 1972).

The relevance of auxotrophic and resistant mutants in the elucidation of the symbiotic pathway is doubtful. We find that virtually any mutational change inflicted on R. *leguminosarum* detracts from its effectiveness on the pea to a greater or lesser degree. In addition, when the bacteria harbour an antibiotic resistance plasmid (RP4) the efficiency of fixation is reduced. It is hard, therefore, to distinguish those mutants with lesions in pathways intimately and specifically involved in the symbiosis from those with only a periferal effect. Clearly, if the plant fails to supply enough of a particular requirement of an auxotroph the Rhizobium may be starved.

The reason for the predelection for using auxotrophs and drug resistant mutants is their relatively simple isolation and manipulation.

Extensive studies from other microbial systems allow us to guess
intelligently at the nature of the metabolic lesions resulting from
such mutations. However, it is possible that mutants of processes
specifically involved in the symbiosis may be qualitatively different.
Selection for mutant phenotypes more closely associated with symbiotic
properties has been attempted, and Kondorosi *et al.* (1973) have described nitrate reductase mutants of Rhizobium that are rendered ineffective. Nitrogenase and nitrate reductase are closely related
enzymes; indeed, the molybdenum-containing subunit of the *Neurospora
crassa* nitrate reductase can substitute for the homologous subunit of
nitrogenase derived from soybean bacteroids (Nason *et al.*, 1971).
However, details of the specific lesions in the ineffective nitrate
reductase deficient mutants are not available. It may be that mutants
lacking nitrate reductase are indeed defective in processes more intimately associated with fixation and with the development of the symbiosis than are auxotrophic or resistant mutants, but this at the present time is largely conjecture.

The culmination of the interaction between Rhizobium and the legume
is, of course, the fixation of nitrogen by nitrogenase. Although
nitrogenase is produced by many prokaryotic microorganisms (Eady and
Postgate, 1974), in the Rhizobiaceae its expression has been detected
only in nodules of legumes or *Trema cannabina*, or in association with
legume callus cultures. Attempts to obtain nitrogen fixation by
rhizobia outside the plant have been made either through mutational
change (Phillips, Howard and Evans, 1973) or by modulation of environmental conditions (Bergersen, 1971; Phillips, Howard and Evans, 1973).
The fact that these attempts have proved unsuccessful indicates that
the legume plays an important role in the expression of this enzyme.
Indeed there has been speculation that the plant may determine at
least one element of the enzyme (Dixon and Cannon, 1975). Using experiments analogous to those employed to establish which partner
makes leghaemoglobin, it has been shown that Fraction 1 of nitrogenase,
at least, is synthesized under the control of genetic material of
bacterial origin (Phillips, Howard and Evans, 1973). These workers
further confirmed that the nitrogenase activity was confined to the
bacteroid fraction of legume nodules. Evidence that the nitrogenase
itself may be coded by a plasmid in *Rhizobium trifolii* has been reported by Dunican and Tierney (1974). They were able to transfer the
ability to fix nitrogen in the free-living state from *R. trifolii* to
a *nif⁻* strain of *Klebsiella aerogenes* and then to *Escherichia coli*

(L.K. Dunican, personal communication). These results, together with those from the callus and *Trema cannabina* work, imply that Rhizobium has the potential for free living fixation, and one might ask why this has not been observed. Perhaps the simplest model is that the plant provides an inducer for the nitrogenase operon (if it exists). However, it is unlikely that a constitutive nitrogenase producing Rhizobium mutant would not have been isolated. Perhaps expression of nitrogenase depends on more than one independent control system, each with its own inducer. Mutationally derived constitutive mutants would then arise at very low frequency, if at all. It may well be that the plant is not intimately involved and that the conditions required for nitrogenase production and expression have not yet been mimicked by experimental manipulation "*in-vitro*".

Implicit in the observation that the Rhizobium codes for the haeme portion of leghaemoglobin and the nitrogenase is that transcription and translation of this information must occur in the rhizobia present in nodules. Therefore, it might be expected that the acetylene-reducing ability of nodules would be greatly reduced when they were exposed to antibiotics which specifically inhibit prokaryote transcription and translation. Our preliminary studies (Table 1) have shown that this indeed occurs when pea plants nodulated with antibiotic-sensitive *R. leguminosarum* strains are treated in this way, though not when nodulated with antibiotic resistant derivatives. This demonstrates that bacterial metabolism is still required for fixation even when the nodule is fully formed and functioning. Analogous experiments to these have been done with Agrobacterium. In contrast to the situation with Rhizobium, crown gall formation was inhibited by rifampicin treatment of the wound only till thirty hours after initial infection (Beiderbeck, 1970) showing that tumour formation by Agrobacterium proceeds in the absence of viable bacteria after an initial stimulus.

To summarize our knowledge of "who does what in the nodule", we can say that the plant provides the structure and nutrition required for the formation of a nodule. Furthermore, it synthesizes the apoprotein of leghaemoglobin which is apparently essential for nitrogenase activity in legume nodules. The rhizobia contribute the haeme moiety of the leghaemoglobin and synthesize the nitrogenase. Both partners must be metabolically active during the development and functioning of the nodule. However, their relative roles in the regulation of the symbiosis are not clear. Further genetic studies

320

Table 1
Acetylene-reduction capacity of whole pea plants (variety Wisconsin Perfection) after exposure to the antibiotics shown.

Nodulating *Rhizobium leguminosarum* strain[*]	Antibiotic	Mean acetylene dependent ethylene production nM/h/g fresh weight (plant)[†]
300	None	85.3
300	Rifampicin	1.1
300 *rif-r*	None	4.9
300 *rif-r*	Rifampicin	12.7
300	Streptomycin	0.87
300 *str-r*	None	7.8
300 *str-r*	Streptomycin	10.0
300	Kanamycin	1.0
300 *kan-r*	None	9.8
300 *kan-r*	Kanamycin	9.8

[*]300 *rif-r*, 300 *str-r* and 300 *kan-r* are respectively rifampicin, streptomycin and kanamycin resistant derivatives of strain 300.

[†]Average from five plants after three days incubation in the presence of antibiotic prior to the assay.

of both rhizobia and legumes are essential to facilitate our understanding of this interaction.

REFERENCES
Becking, J.H. (1970). Plant-endophyte symbiosis in non-leguminous plants. *Plant and Soil.* 32, 611-654.

Beiderbeck, R. (1970). Untersuchungen an Crown-Gall IV. Rifampicin und ein resistenter Klon von *Agrobacterium tumefaciens* bei der Tumorinduktion. *Zeitung für Naturforschung.* 25b, 1458-1460.

Broughton, W.J., and Dilworth, M.J. (1971). Control of leghaemoglobin synthesis in snake beans. *Biochemical Journal.* 125, 1075-1080.

Cutting, J.A., and Schulman, H.M. (1969). The site of haeme synthesis in soybean root nodules. *Biochimica et Biophysica Acta.* 192, 486-493

Cutting, J.A., and Schulman, H.M. (1971). The biogenesis of leghaemoglobin. The determinant in the Rhizobium-legume symbiosis for leghaemoglobin specificity. *Biochimica et Biophysica Acta.* 229, 58-62.

Dart, P.J. (1974). The infection process. In *Biological Nitrogen*

Fixation, Edited by A. Quispel. Amsterdam: North Holland.

Dénarié, J., Truchet, G., and Bergeron, B. (1975). Effects of some mutations on symbiotic properties of Rhizobium. In *Symbiotic Nitrogen Fixation in Plants,* Edited by P.S. Nutman. I.B.P. Volume 7. Cambridge: Cambridge University Press.

Dilworth, M.J. (1969). The plant as the genetic determinant of leghaemoglobin production in the legume root nodule. *Biochimica et Biophysica Acta.* 184, 432-441.

Dixon, R.O.D. (1969). Rhizobia (with particular reference to relationships with host plants). *Annual Reviews of Microbiology.* 23, 137-158.

Dixon, R.A., and Cannon, F.C. (1975). Recent advances in the genetics of nitrogen fixation. In *Sybiotic Nitrogen Fixation in Plants,* Edited by P.S. Nutman. I.B.P. Volume 7. Cambridge: Cambridge University Press.

Dunican, L.K., and Tierney, A.B. (1974). Genetic transfer of nitrogen fixation from *Rhizobium trifolii* to *Klebsiella aerogenes. Biochemical and Biophysical Research Communications.* 57, 62-72.

Eady, R.R., and Postgate, J.R. (1974). Nitrogenase. *Nature, London.* 249, 805-810.

Elkan, G.H. (1961). A nodulation-inhibiting root excretion from a non-nodulating soybean strain. *Canadian Journal of Microbiology.* 7, 851-856.

Gibbins, A.M., and Gregory, K.F. (1972). Relatedness among Rhizobium and Agrobacterium species determined by three methods of nucleic acid hybridization. *Journal of Bacteriology.* 111, 129-141.

Graham, P.H. (1964). The application of computer techniques to the taxonomy of the root-nodule bacteria of legumes. *Journal of General Microbiology.* 35, 511-517.

Hendry, G.S., and Jordan, D.C. (1969). Ineffectivness of viomycin-resistant mutants of *Rhizobium meliloti. Canadian Journal of Microbiology.* 15, 671-675.

Kleczkowska, J. (1950). A study of phage-resistant mutants of *Rhizobium trifolii. Journal of General Microbiology.* 4, 298-310.

Kondorosi, Á., Barabás, I., Sváb, Z., Orosz, L., Sik, T., and Hotchkiss, R.D. (1973). Evidence for common genetic determinants of nitrogenase and nitrate reductase in *Rhizobium meliloti. Nature, New Biology.* 246, 153-154.

Kowalski, M., and Dénarié, J. (1972). Transduction d'un gene contrôlant l'expression de la fixation de l'azote chez *Rhizobium meliloti.*

322

Comptes Rendus de l'Académie des Sciences, Paris, Série D. 275, 141-144.

MacKenzie, C.R., and Jordan, D.C. (1970). Cell wall phospholipid and viomycin resistance in *Rhizobium meliloti*. *Biochemical and Biophysical Research Communications*. 40, 1008-1012.

Nason, A., Lee, K., Pan, S., Ketchum, P.A., Lamberti, A., and Devries, J. (1971). *In vitro* formation of assimilatory reduced nicotinamide adenine dinucleotide phosphate: nitrate reductase from a *Neurospora* mutant and a component of molybdenum-enzymes. *Proceedings of the National Academy of Sciences, U.S.A.* 68, 3242-3246.

Nutman, P.S. (1955). The influence of the legume in root-nodule symbiosis. A comparative study of host determinants and functions. *Biological Reviews*. 31, 109-151.

Nutman, P.S. (1969). Genetics of symbiosis and nitrogen fixation in legumes. *Proceedings of the Royal Society. Series B.* 172, 417-437.

Pankhurst, C.E. (1974). Ineffective *Rhizobium trifolii* mutants examined by immune-diffusion, gel-electrophoresis and electron microscopy. *Journal of General Microbiology*. 82, 405-413.

Pankhurst, C.E., Schwinghamer, E.A., and Bergersen, F.J. (1972). The structure and acetylene-reducing activity of root nodules formed by a riboflavin-requiring mutant of *Rhizobium trifolii*. *Journal of General Microbiology*. 70, 161-177.

Phillips, D.A. (1971). A cotyledonary inhibitor of root nodulation in *Pisum sativum*. *Physiologia Plantarum*. 25, 482-487.

Scherrer, A., and Dénarié, J. (1971). Symbiotic properties of some auxotrophic mutants of *Rhizobium meliloti* and their prototrophic revertants. *Plant and Soil, Special Volume*. 39-45.

Schwinghamer, E.A. (1964). Association between antibiotic resistance and ineffectiveness in mutant strains of Rhizobium spp. *Canadian Journal of Microbiology*. 10, 221-233.

Schwinghamer, E.A. (1969). Mutation to auxotrophy and prototrophy as related to symbiotic effectiveness in *Rhizobium leguminosarum* and *R. trifolii*. *Canadian Journal of Microbiology*. 15, 611-622.

Schwinghamer, E.A. (1970). Requirement for riboflavin for effective symbiosis on clover by an auxotrophic mutant strain of *Rhizobium trifolii*. *Australian Journal of Biological Sciences*. 23, 1187-1196.

Staphorst, J.L., and Strijdom, B.W. (1971). Infectivity and effectiveness of colony isolates of ineffective glycine resistant *Rhizobium meliloti* strains. *Phytophylactica* 3, 131-136.

Tanner, J.W., and Anderson, I.C. (1963). Investigations on non-nodulating and nodulating soybean strains. *Canadian Journal of Plant*

Science. 43, 542-546.

Trinick, M.J. (1973). Symbiosis between Rhizobium and the non-legume, *Trema aspera. Nature, London.* 244, 459-460.

Verna, D.P.S., Nash, D.T., and Schulman, H.M. (1974). Isolation and *in vitro* translation of soybean leghaemoglobin mRNA. *Nature, London.* 257, 74-77.

Williams, L.F., and Lynch, D.L. (1954). Inheritance of non-nodulating character in soybean. *Agronomy.* 46, 28-29.

Wittenberg, J.B., Bergerson, F.J., Appleby, C.A., and Turner, G.L. (1974). Facilitated Oxygen Diffusion. *Journal of Biological Chemistry.* 249, 4057-4066.

Yu, K.K.-Y., and Jordan, D.C. (1971). Cation content and cation-exchange capacity of intact cells and cell envelopes of viomycin-sensitive and -resistant strains of *Rhizobium meliloti. Canadian Journal of Microbiology.* 17, 1283-1286.

SYMPOSIUM PARTICIPANTS

ALFERMANN, W., Lehrstuhl für Pharmazentische Biologie, D-74 Tübingen, Auf der Morganstelle 8, Germany.

AUSUBEL, F.M., Department of Botany, University of Leicester, Leicester, LE1 7RH, U.K.

AVANZI, S., Institute of Botany, The University of Pisa, 56100 Pisa, Italy.

BAGNI, N., Institute of Botany, University of Bologna, Via Irnerio 42, 40126 Bologna, Italy.

BAKHUISEN, C.E.G.C., State University Leiden, Department of Biochemistry, Wassenaarseweg 64, Leiden, The Netherlands.

BANCROFT, J.B., Department of Plant Sciences, University of Western Ontario, London, Ontario, Canada.

BARABÁS, I., Institute of Genetics, Biological Research Center, Hungarian Academy of Sciences, 6701 Szeged, P.O.B. 521, Hungary.

BARBARA, D.J., Department of Virology, Medical School, University of Birmingham, Birmingham 15, U.K.

BARNETT, A., The University of Birmingham, Department of Microbiology, P.O. Box 363, Birmingham B15 2TT, U.K.

BEACHY, R.N. Department of Plant Pathology, Cornell University, Ithaca, New York 14850, U.S.A.

BEEFTINK, F., Institute of Molecular Biology, Free University of Brussels, Paardenstraat 65, 1670 St. Genesius-Rode, Belgium.

BEHKI, R.M., Department of Radiobiology, Section of Cellular Biochemistry, C.E.N./S.C.K., B-2400 Mol, Belgium.

BERINGER, J.E., Department of Genetics, John Innes Institute, Colney Lane, Norwich, NR4 7UH, U.K.

BERRY, C.H.J., School of Biological Sciences, University of East Anglia, University Plain, Norwich, NOR 88C, U.K.

BONGA, J.M., Maritime Forest Research Center, Box 4000 Fredericton, New Brunswick, Canada.

BOURGIN, J.P., Physiologie Végétale, CNRA, Route de St. Cyr, 78000 Versailles, France.

BOVÉ, J.M., Station de Physiologie et Biochimie Végétales "La Grande Ferrade", 33140 Pont-de-la-Maye, France.

BRAIN, K.R., Welsh School of Pharmacy, U.W.I.S.T., Cardiff, CF1 3NU, U.K.

BRIGHT, S.W.J., University of Cambridge, Department of Biochemistry, Tennis Court Road, Cambridge, CB2 1QW, U.K.

BUIATTI, M., Instituto di Genetica, Via Matteotti 1/A, University of Pisa, 56100 Pisa, Italy.

325

BURR, K.W., Glaxo Laboratories Ltd., Ulverston, Lancashire, LA12 9DR, U.K.

BURROWS, W.J., Shell Research Limited, Research Centre, Sittingbourne, Kent, U.K.

CALLOW, J.A., Department of Plant Sciences, Baines Wing, University of Leeds, Leeds 2, U.K.

CANNON, F., A.R.C. Unit of Nitrogen Fixation, University of Sussex, Brighton, BN1 9QJ, U.K.

CAPESIUS, I., 69 Heidelberg, Hofmeisterweg 4, Botanisches Institut, Germany.

CARTWRIGHT, T.E., 336 Clapp Hall, Biophysics and Microbiology Department, University of Pittsburgh, Pittsburgh, Pennsylvania 15260, U.S.A.

CATARINO, F.M., Instituto Botanico, Faculdade de Ciencias, Lisboa, Portugal.

CHALEFF, R.S., Department of Applied Genetics, John Innes Institute, Colney Lane, Norwich, NR4 7UH, U.K.

CHATER, K.R., Department of Genetics, John Innes Institute, Colney Lane, Norwich NR4 7UH, U.K.

CHILTON, M.D., Department of Microbiology, University of Washington, Seattle, Washington 98195, U.S.A.

CIONINI, P.G., Instituto di Genetica, Via Matteotti 1/A, University of Pisa, 56100 Pisa, Italy.

COCKING, E.C., Department of Botany, University of Nottingham, University Park, Nottingham, NG7 2RD, U.K.

CODDINGTON, A., School of Biological Sciences, University of East Anglia, University Plain, Norwich, NOR 88C, U.K.

CONTI, G.G., Unita di Ricerca Virus e Virosi Plante, C.N.R., c/o Instituto Patalogia Vegetale, Via Celoria 2, 20133 Milano, Italy.

COOPER, D., Department of Genetics, John Innes Institute, Colney Lane, Norwich, NR4 7UH, U.K.

COUTTS, R.H.A., The Department of Microbiology, University of Birmingham, Box 363, Birmingham, B15 2TT, U.K.

CREMONINI, R., Instituto di Genetica, Via Matteotti 1/A, 56100 Pisa, Italy.

CULLIS, C.A., Department of Applied Genetics, John Innes Institute, Colney Lane, Norwich, NR4 7UH, U.K.

DANIELS, M.J., Department of Genetics, John Innes Institute, Colney Lane, Norwich, NR4 7UH, U.K.

DARBY, G.K., Department of Virology, Medical School, University of Birmingham, Birmingham, 15, U.K.

DAVEY, M.R., Botany Department, University Park, Nottingham, NG7 2RD, U.K.

DAVIES, D.R., Department of Applied Genetics, John Innes Institute, Colney Lane, Norwich, NR4 7UH, U.K.

DAWSON, J.R.O., Department of Virus Research, John Innes Institute, Colney Lane, Norwich, NR4 7UH, U.K.

DE GROOT, B., Genetisch Laboratorium, Kaisersstr. 63, Leiden, The Netherlands.

DE GUZMAN, E.V., Department of Agronomy, College of Agriculture, University of Philippines at Los Baños, College Laguna, Philippinnes.

DIENER, T.O., Plant Virology Laboratory, U.S.D.A. Agricultural Research Service, Plant Protection Institute, Beltsville, Maryland 20703, U.S.A.

DIXON, R.A., A.R.C. Unit of Nitrogen Fixation, University of Sussex, Brighton, BN1 9QJ, U.K.

DRUMMOND, M.H., Department of Genetics, Leeds University, Leeds, 2, U.K.

D'SOUZA, S., University of Aberdeen, Department of Genetics, 2 Tilly-drone Avenue, Aberdeen, U.K.

DUBOIS, J., Ministere de l'Agriculture, Station d'Amelioration des Plantes, 5800 Gembloux, Rue du Bordia 4, Belgium.

DUNN, D.B., Department of Virus Research, John Innes Institute, Colney Lane, Norwich, NR4 7UH, U.K.

DUNWELL, J.M., Department of Applied Genetics, John Innes Institute, Colney Lane, Norwich, NR4 7UH, U.K.

DURANTE, M., Instituto di Genetica, Via Matteotti 1/A, 56100 Pisa, Italy.

DURZAN, D.J., Forest Ecology Research Institute, Canadian Forestry Service, Ottawa, K1A OW5, Canada.

ELLIOTT, T.J., Glasshouse Crops Research Institute, Worthing Road, Littlehampton, Sussex, U.K.

ERIKSSEN, T., Institute of Physiological Botany, S-75121, Uppsala, Sweden.

EVANS, P.K., Botany Department, University of Nottingham, Nottingham, NG7 2RD, U.K.

FINCHAM, J.R.S., Department of Genetics, The University, Leeds, LS2 9JT, U.K.

FLAVELL, R.B., Cytogenetics Department, Plant Breeding Institute, Maris Lane, Trumpington, Cambridge, CB2 2LQ, U.K.

FORAGE, A.J., Glaxo Laboratories Ltd., Ulverston, Cumbria, LA12 9DR, U.K.

FRACASSINI, D.S., Institute of Botany, University of Bologna, Via Irnerio 42, 40126 Bologna, Italy.

FURNER, I., School of Biological Sciences, The University of Sussex, Biology Building, Falmer, Brighton, Sussex, U.K.

GALBRAITH, D.W., University of Cambridge, Department of Biochemistry, Tennis Court Road, Cambridge, U.K.

GALUN, E., The Weizmann Institute of Science, Rehovot, Israel.

GAMBORG, O.L., National Research Council Canada, Prairie Regional Laboratory, 110 Gymnasium Road, University Campus, Saskatoon, Sask. Canada, S7N OW9. NRCC No. 14385.

GERI, C., c/o Instituto di Genetica, Viale Matteotti 1/A, 56100 Pisa, Italy.

GILES, K.L., Plant Physiology Division, D.S.I.R., Private Bag, Palmerston North, New Zealand.

GIORGI, L., Instituto di Genetica, Via Matteotti 1/A, University of Pisa, 56100 Pisa, Italy.

GOODMAN, R.M., Department of Virus Research, John Innes Institute, Colney Lane, Norwich, NR4 7UH, U.K.

GRAFE, R., Zentralinstitut f. Genetik u. Kulturpflanzenforschung der AdW der DDR, 4325 Gatersleben, Germany.

GRESSEL, J., Weizmann Institute of Science, Department of Plant Genetics, Rehovot, Israel.

GRIERSON, D., Department of Physiology and Environmental Studies, University of Nottingham, School of Agriculture, Sutton Bonington, Loughborough, LE12 5RD, U.K.

GUILLÉ, E., Laboratoire de Biologie Moléculaire végétale, Universite Paris-Sud, Bat. 430, Centre d'Orsay, Orsay 91405, France.

HANSCHE, P.E., University of California, Davis, 1053 Wickson Hall, Davis, California 95616, U.S.A.

HARRISON, B.D., Scottish Horticultural Research Institute, Invergowrie, Dundee, DD2 5DA, U.K.

HEYN, R.F., Department of Biochemistry, University of Leiden, Wassenaarseweg 64, Leiden, The Netherlands.

HOLL, B., N.R.C. Prairie Regional Laboratory, Saskatoon, Canada.

HOPWOOD, D.A., Department of Genetics, John Innes Institute, Colney Lane, Norwich, NR4 7UH, U.K.

HORNE, R.W., Department of Ultrastructural Studies, John Innes Institute, Colney Lane, Norwich, NR4 7UH, U.K.

HUBER, R., Molecular Biology Department, Agricultural University, De Dreyen 11, Wageningen, The Netherlands.

HUGHES, M.A., Genetics Department, Ridley Building, The University, Newcastle-Upon-Tyne, NE1 7RU, U.K.

HUTCHINSON, J.B., Huntingfield, Huntingdon Road, Cambridge, U.K.

IBRAHIM, M.A.K., School of Biological Sciences, University of East Anglia, University Plain, Norwich, NOR 88C, U.K.

INGLE, J., University of Edinburgh, Department of Botany, The King's Buildings, Mayfield Road, Edinburgh, EH9 3JH, U.K.

JACOBS, M., Vrije Universiteit Brussel, I.M.B. Paardenstraat 65, 1640 St. Genesius-Rode (B), Belgium.

JAWORSKI, E.G., Monsanto Commercial Products Co., 800 N. Lindbergh Boulevard, St. Louis, Missouri, 63166, U.S.A.

JOHNSTON, A.W.B., Department of Genetics, John Innes Institute, Colney Lane, Norwich, NR4 7UH, U.K.

JONES, M., School of Biological Sciences, University of East Anglia, University Plain, Norwich, NOR 88C, U.K.

KENNEDY, C., A.R.C. Unit of Nitrogen Fixation, University of Sussex, Falmer, Brighton, Sussex.

KLEINHOFS, A., Department of Radiobiology, S.C.K./C.E.N., B-2400 Mol, Belgium.

KOHLENBACH, H.W., FB Biologie-Botanik-der Universität, 6 Frankfurt Am Main, Siesmayerstr. 70 BRD, Germany.

KONDOROSI, A., A.R.C. Unit of Nitrogen Fixation, University of Sussex, Brighton, BN1 9QJ, U.K.

KUBO, S., c/o Scottish Horticultural Research Institute, Invergowrie, Dundee, DD2 5DA, U.K.

KUMMERT, J., Laboratoire de Pathologie Végétale, Faculté des Sciences Agronomiques, 8, Avenue Marechal Juin, 5800 Gembloux, Belgium.

LA COUR, L.F., School of Biological Sciences, University of East Anglia, University Plain, Norwich, NOR 88C, U.K.

LÁZÁR, G., Biological Research Center, Hungarian Academy of Sciences, Szeged, P.O.B. 521, 6701, Hungary.

LEDOUX, L., Centre d'Etude de l'Energie Nucleaire, C.E.N./S.C.K., Boertang 200, B-2400 Mol, Belgium.

LEE, P.E., Department of Biology, Carleton University, Ottawa, Ontario, K1S 5B6, Canada.

LOENING, U.E., Department of Zoology, University of Edinburgh, The King's Buildings, Mayfield Road, Edinburgh, EH9 3JT, U.K.

LESLEY, S.M., Canadian Department of Agriculture, Research Branch, Chemical and Biological Research Institute, Ottawa, Ontario, K1A 0C6, Canada.

LINDLEY, H., Flat 5, Sandhill Court, Sandhill Lane, Leeds, 17, U.K.

LUI, M-C., Taiwan Sugar Research Institute, 54 Sheng Chan Road, Tainan, Taiwan.

MAGRUM, L.J., Department of Ultrastructural Studies, John Innes Institute, Colney Lane, Norwich, NR4 7UH, U.K.

MALIGA, P., Institute of Plant Physiology, 6701 Szeged, P.O.B. 521, Hungary.

MAPELLI, S., Laboratorio Virus e Biosintesi Vegetali, C.N.R., Via Celoria 2, 20133 Milano, Italy.

MARETZKI, A., c/o Haiwaiin Sugar Planters' Association, 1527 Keeaumoku Street, Honolulu, Haiwaii 96821.

MARKHAM, R., Director, John Innes Institute, Colney Lane, Norwich, NR4 7UH, U.K.

MARTINI, G., Laboratorio di Mutagenesi e Differenziamento, Viale Matteotti 1/A, 56100 Pisa, Italy.

MÁRTON, L., Institute of Plant Physiology, 6701 Szeged, P.O.B. 521, Hungary.

MASSARI, M.G., Corso Lodi 15, Milano, Italy.

MAURY, Y., Chargé du Recherches, CNRA Station Centrale de Pathologie Végétale, Etoile de Choisy, Route de Saint-Cyr, 7800 Versailles, France.

MERRICK, M.J., Department of Genetics, John Innes Institute, Colney Lane, Norwich, NR4 7UH, U.K.

McCARTHY, D., Microbiology and Plant Biology Department, Queen Mary College, Mile End Road, London, E.1., U.K.

MELCHERS, G., Director, Max-Planck-Institut für Biologie, 74-Tübingen, Correnstrasse 41, Germany.

MIELICKE, G., 74-Tübingen, Auf der Morgenstelle, Institut für Biologie (Lab Dr. Sander), Germany.

MIFLIN, B.J., Department of Biochemistry, Rothamsted Experimental Station, Harpenden, Hertfordshire, U.K.

MINSON, A.C., Department of Virology, Medical School, University of Birmingham, Birmingham, 15, U.K.

MORRIS, R.O., Department of Agricultural Chemistry, Oregon State University, Corvallis, Oregon, 97331, U.S.A.

MOUCHES, C., INRA, Station de Physiologie et de Biochimie Végétales "La Grande Ferrade", 33 Pont-de-la-Maye, France.

MUNGALL, J., University of Edinburgh, Department of Botany, The King's Buildings, Mayfield Road, Edinburgh, EH9 3JH, U.K.

MURAKISHI, H.H., Department of Botany and Plant Pathology, Michigan State University, East Lansing, Michigan, 48823, U.S.A.

NAGL, W., Department of Biology, Division of Cytology, The University, Pfaffenbergstr. 95, D-675 Kaiserslautern, Germany.

NEGRUTIU, I., Institute of Molecular Biology, Free University of Brussels, Paardenstraat 65, 1640 St. Genesius Rode, Belgium.

NUTI RONCHI, V., Laboratorio di Mutagenesi e Differenziamento, Viale Matteotti 1/A, 56100 Pisa, Italy.

OXELFELT, P., Department of Plant Pathology and Entomology, Agricultural College of Sweden, S-75007, Uppsala 7, Sweden.

PAIN, A., Department of Genetics, The University of Liverpool, Brownlow Street, P.O. Box 147, Liverpool, L69, 3BX, U.K.

PARENTI, R., Laboratorio di Mutagenesi e Differenziamento, C.N.R., Via Matteotti 1/A, 56100 Pisa, Italy.

PLA, J., UNESCO Chief Technical Adviser, Biological Research Center H-6701 Szeged, P.O.B. 521, Hungary.

PONTECORVO, G., Imperial Cancer Research Fund Laboratories, P.O. Box 123, Lincoln's Inn Fields, London, WC2A 3PX, U.K.

POSTGATE, J.R., Department of Microbiology, The University of Sussex, Falmer, Brighton, Sussex, U.K.

POTRYKUS, I., Max-Planck-Institut fur Pflanzengenetik, Projektforschung Haploids, D-6802 Ladenberg/Heidelberg, Germany.

RABSON, R., Plant Breeding and Genetics Section, International Atomic Energy Agency, P.O.B. 590, A-1011 Wien, Austria.

REDEI, G.P., University of Missouri-Columbia, Department of Agronomy, 204 Curtis Hall, Columbia, Missouri 65201, U.S.A.

REES, M.W., Department of Virus Research, John Innes Institute, Colney Lane, Norwich, NR4 7UH, U.K.

REINERT, J., Institut für Pflanzenphysiologie and Zellbiologie, 1 Berlin 33, Königin-Luise-Str. 12-16a, Germany.

RENAUDIN, J., I.N.R.A., Station de Physiologie et de Biochimie Végétales, "La Grande Ferrade", 33 Pont-de-la-Maye, France.

RIMPAU, J., Plant Breeding Institute, Maris Lane, Trumpington, Cambridge, U.K.

ROBERTS, K., Department of Ultrastructural Studies, John Innes Institute, Colney Lane, Norwich, NR4 7UH, U.K.

ROBINSON, D.J., Scottish Horticultural Research Institute, Invergowrie, Dundee, DD2 5DA, U.K.

ROMANI, R.J., Department of Pomology, University of California, Davis, California, 95616, U.S.A.

RÖRSCH, A., Department of Biochemistry, State University of Leiden, Wassenaarseweg 64, Leiden, The Netherlands.

RÖTZER, R., 7407 Rottenburg - 6, Hauptstr. 36, Germany.

SACRISTÁN, M-D., Max-Planck-Institut für Biologie, 7400 Tübingen, Correnstrasse 41, Germany.

SAKAI, F., Department of Ultrastructural Studies, John Innes Institute, Colney Lane, Norwich, NR4 7UH, U.K.

SÄNGER, H.L., Arbeitsgruppe Pflanzenvirologie, Fachbereich Angewandte Biologie, Justus Liebig-Universität, D 6300 Giessen, Frankfurter Strasse 107, Germany.

SARTIRANA, M.L., Department of Zoology, University of Edinburgh, West Mains Road, Edinburgh, EH9 3JH, U.K.

SCHILDE-REUTSCHLER, L., Max-Planck-Institut für Biologie, 74 Tübingen, Correnstrasse 41, Germany.

SCHILPEROORT, R.A., Biochemisch Laboratorium, Rijksuniversiteit Leiden, Wassenaarseweg 64, Leiden, The Netherlands.

SEMAL, J., Laboratoire de Pathologie Végétale, Faculté des Sciences Agronomiques de l'Etat, 5800 Gambloux, Belgium.

SHOULDER, A., The University of Birmingham, Department of Virology, The Medical School, Birmingham, B15 2TJ, U.K.

SIEGEL, A., Biology Department, Wayne State University, College of Liberal Arts, Detroit, Michigan, 48202, U.S.A.

SISSÖEFF, I., Laboratoire de Biologie Moléculaire, Bat 430, Université Paris-Sud, 91405 Orsay, France.

SMITH, H., Department of Physiology and Environmental Studies, School of Agriculture, Sutton Bonington, Loughborough, LE12 5RD, U.K.

STERN, H., Department of Biology, University of California, San Diego Campus, P.O.B. 109 La Jolla, California, 92037, U.S.A.

SUNG, Z.R., 56.444 Massachusetts Institute of Technology, Cambridge, Massachusetts, 02139, U.S.A.

SZIDONYA-BREZNOVITS, A., c/o Mr. Pla, Chief Technical Adviser, Biological Research Center, Hungarian Academy of Sciences, Box 521, Szeged, Hungary.

SZWEYKOWSKA, A., Institute of Biology, The University, Stalingradzka, 14, 61-713, Poznan, Poland.

TABATA, M., Faculty of Pharmaceutical Sciences, Kyoto University, Sakyo-Ku, Kyoto, Japan.

TAKEBE, I., Institute for Plant Virus Research, 959 Aobacho, Chiba 280, Japan.

TEMPÉ, J., Centre National de Recherches Agronomiques, Route de St. Cyr, 7800 Versailles, France.

THORPE, T.A., Department of Biology, University of Calgary, Calgary, Alberta, Canada, T2N 1N4.

TUBB, R.S., A.R.C. Unit of Nitrogen Fixation, University of Sussex, Brighton, BN1 9QJ, U.K.

UETAKE, H., Institute for Virus Research, Kyoto University, Sakyo-Ku, Kyoto, Japan.

UPADHYA, M.D., Max-Planck Institut für Biologie, Corrensstr. 41, 74 Tübingen, Germany.

VEGETTI, G., Unita di Ricerca Virus Evirosi Plante, C.N.R., c/o Istito Patologia Vegetale, Via Celoria 2, 20133 Milano, Italy.

WALLACE, H., Department of Genetics, University of Birmingham, P.O. Box 363, Birmingham, B15 2TT, U.K.

WARREN-WILSON, M.W., University of Reading, Department of Microbiology, Wessex Hall, Whiteknights Road, Reading, RG6 2BQ, U.K.

WATTS, J.W., Department of Ultrastructural Studies, John Innes Institute, Colney Lane, Norwich, NR4 7UH, U.K.

WHITE, R.F., Plant Pathology Department, Rothamsted Experimental Station, Harpenden, Hertfordshire, U.K.

WILDY, P., The University of Birmingham, Department of Virology, The Medical School, Birmingham, B15 2TJ, U.K.

WILDON, D., School of Biological Sciences, University of East Anglia, University Plain, Norwich, NOR 88C, U.K.

WILLIAMS M.H., The Welsh School of Pharmacy, U.W.I.S.T., King Edward VII Avenue, Cardiff CF1 3NU, U.K.

WOOD, K.R., University of Birmingham, Department of Microbiology, P.O. Box 363, Birmingham, B15 2TT, U.K.

YAMADA, Y., Department of Agricultural Chemistry, Kyoto University, Sakyo, Kyoto 606, Japan.

ZAITLIN, M., Department of Plant Pathology, Cornell University, Ithaca, New York, 14850, U.S.A.

ZIMMERN, D., MRC Laboratory of Molecular Biology, University Postgraduate Medical School, Hills Road, Cambridge, CB2 2QH, U.K.

INDEX

Protoplasts
   use of in plant science, 170
   viral protein, 155
   Zea mays, 174
PS8
   phage, 253
   phage in crown gall tumour tissue, 298
Pseudomonas patida
   5-fluorotryptophan-resistant strains, 286
Pseudomonas tabaci
   toxin, 198

Repeated nucleotide sequences -
   in rye chromosomes, 57
Rhizobium, 313
   effectiveness, 316
   pleomorphic bacteriods, 316
   resistance to antibiotics, 316
Rhizobium meliloti
   leucine auxotroph, 317
Ribonucleic acid
   see RNA
Ribosomal RNA, 53
Ribosomal RNA genes, 33
Rice
   callus culture, 201
Rice dwarf virus
   RNA, 104
RNA
   -BMV, 137
   -CCMV, 137
   denaturation with dimethylsulphoxide, 104
   double-stranded, 104
   electrophoretic separation, 102
   high molecular weight, 64
   nucleotide composition of, 106
   ribosomal, 16, 53
   rice dwarf virus, 104
   single-stranded CCMV-WT, 146
   tobacco mosaic virus, 101, 137
   tobacco mosaic virus pseudovirion, 18

348

RNA
   viral double-stranded, 154
   virus-related low molecular weight, 153
Root-nodules
   of legumes, 313
Rutgers tomato plants, 230
Rye, 53, 55

Sänger, H.L., 229
Satellite DNA, 42
Schilperoort, R.A., 253
Schistocerca gregaria
   DNA in, 38
Secale cereale, 53
Secale vavilovi, 59
Sendai virus
   in the fusion of human cells, 7
Sex
   alternatives to, 1
Siegel, A., 15
Sinapis alba, 81
Sodium azide
   as inhibitor of protein synthesis, 126
   as metabolic inhibitor, 140
Somatic cells
   genetics by means of, 1
Somatic cell hybridisation
   -human, 3
Somatic hybridisation, 161, 181
'Spacer' sequences of RNA, 53
Stern, H., 63
Stormont cirrus
   DNA changes in the flax, 27
Synteny, 5
Sunflower
   callus DNA, 272
   tumour DNA, 272
Sycamore cell
   β-galactosidase activity, 94
   cultures. 92
   growth in glucose, 95